AF797481

YOU NEVER DIE

YOU NEVER DIE

A Scientific Case for Immortality

MARK DOUGLAS

Copyright © 2026 by Mark Douglas

All rights reserved. No part of this publication may be reproduced, distributed, or transmitted in any form or by any means, including photocopying, recording, or other electronic or mechanical methods, without the prior written permission of the publisher, except in the case of brief quotations embodied in critical reviews and certain other noncommercial uses permitted by copyright law.

Published by Zasder Publishing

Cover and interior design: Karolina Kruk-Umięcka

ISBN: 978-1-972209-01-1 (paperback)

ISBN: 978-1-972209-04-2 (ebook)

First Edition

Contents

PREFACE:

The Question That Won't Go Away

You are going to die.

Your heart will stop. Your neurons will fire for the last time. The people who love you will put your body in the ground or scatter your ashes, and the world will continue without you. This is the one fact about your life that you can be absolutely certain of.

Or is it?

For most of human history, people have handled this certainty in one of two ways. The religious say: don't worry, there's an afterlife: a soul, a heaven, a reincarnation, something beyond the grave. The materialists say: sorry, when you're dead, you're dead. There is nothing else. Learn to accept it.

Both sides, I believe, are wrong.

Not because the truth lies somewhere in the mystical middle, but because both sides have failed to think carefully enough about what consciousness actually **is** and what the laws of physics actually allow. When you do think carefully — rigorously, scientifically, without any appeal to the supernatural — you arrive at a conclusion that may shock you:

You cannot permanently die.

Not in the way you think. Not as a conscious being.

This is not wishful thinking or religion dressed up in a lab coat. It is the logical consequence of three premises drawn from mainstream neuroscience and physics. The problem is that almost nobody has followed these premises to their natural conclusion.

That's what this book does.

I have spent decades developing what I call **Ixperiencit Theory**, a scientific framework for understanding consciousness, identity, and immortality that requires nothing supernatural. No souls. No gods. No supernatural mechanisms. Just the structure and functioning of matter and energy, operating according to the laws of physics as we understand them, and yet, as you will see, the implications go further than any heaven that religion has imagined.

Let me be clear about what this book is and what it is not. It is not the complete theory. The full Ixperiencit Theory framework, developed over decades, continues to grow. This book is an invitation, a way to show you the core ideas, walk you through the logic, explore the thought experiments, and let you see for yourself why the implications are staggering. I have tried to write it the way I would explain these ideas to a smart friend over a long dinner: clearly, honestly, with real examples and vivid analogies, and without shying away from the hard questions.

If you are religious, this book will show you something better than what your religion promises, grounded in science rather than faith.

If you are an atheist, this book will show you that accepting materialism does not require accepting permanent death.

If you are a scientist, this book will challenge you to follow your own assumptions to their logical end.

If you are a philosopher, this book will give you new tools for thinking about the oldest question there is.

And if you are simply someone who has ever lain awake at night wondering what happens when you die — if you have ever felt that cold knot of dread in your stomach when you really contemplate oblivion — this book is for you. Because the answer is far more surprising than you have been led to believe.

How This Book Works

The argument unfolds in seven stages:

Part One destabilizes your intuitions about death and identity, and establishes the three scientific premises on which everything else rests.

Part Two builds the theoretical framework: the vocabulary and concepts you need to think precisely about consciousness, identity, and survival.

Part Three stress-tests the theory with thought experiments, pushing the premises to their logical extremes and seeing whether they hold. Part Three will raise objections in your mind. Part Four takes those objections seriously.

Part Four confronts the strongest objections head-on (Parfit on causal continuity, Chalmers on the hard problem, the measure problem from physics) and responds to each honestly.

Part Five draws the conclusions: what superimmortality actually means, what forms it takes, and how different versions of your consciousness relate to each other.

Part Six places these conclusions in the wider context of religion, philosophy, science, and technology.

Part Seven explores the implications for ethics, grief, the future of science, and how you live your life right now.

Each part builds on the previous one. If you find yourself skeptical in Part Two, stay with it until Part Four, because the objections you're forming are addressed there. If you find the thought experiments too abstract, the neuroscience chapter in Part Two grounds them in laboratory reality.

Let's begin.

PART ONE: THE PROBLEM

CHAPTER 1

Death Is Not What You Think

Here is a question almost no one asks: What exactly ends when you die?

Most people assume the answer is obvious. Everything ends. You cease to exist. But let's be more precise, because precision matters enormously here.

When you die, your body stops functioning. The heart stops pumping blood, the lungs stop drawing air, and the brain stops processing information. Your atoms begin to disperse; they return to the earth, the air, the water. Within a few years, virtually every atom that was once "you" is somewhere else entirely.

The Existing Answers

Humanity has proposed five answers to the question of what happens when you die. None is fully satisfying.

Reincarnation. The oldest answer: a soul leaves your body at death and enters a new one, whether human, animal, or divine. Some variants (the "born-again" or déjà vu theory) focus on memories and abilities that seem to carry over between lives. The scientific problem is fundamental: there is no evidence for any soul-like entity that could travel between bodies, and reincarnation into a radically different organism implies consciousness is not produced by brain structure, contradicting strong evidence that it is.

Heaven and Hell. The supernatural version requires otherworldly realms, divine judges, and souls, none of which have evidential support. It answers the emotional need for justice and reunion but demands belief in mechanisms that lie entirely outside what science can observe or test.

Death Is an Illusion. This view holds that consciousness is fundamental and bodies are constructs within it, rather than consciousness being a product of bodies. There is an interesting kernel here (as this book will argue, permanent experiential death is an illusion) but the theory typically denies the reality of bodily death altogether, which is plainly false. Bodies really do stop functioning.

Death Is Final. The standard materialist position: when the brain dies, consciousness ends, permanently. Of the traditional theories, this one fits the scientific evidence best. It correctly identifies consciousness as a product of brain functioning and correctly rejects supernatural concepts. But it draws the wrong conclusion from the right premises, conflating the death of a body with the death of a consciousness, assuming your experience is tied to one specific collection of matter.

This book proposes a different answer, one built entirely on premises that science already accepts. And it begins with a fact that sounds mundane but turns out to be explosive.

But here is the critical insight that almost everyone misses: **those atoms were never really "you" in the first place.**

You already know this, if you think about it. The atoms in your body are constantly being replaced. You shed skin cells, breathe out carbon, excrete waste, and take in new atoms from food, water, and air. Over the course of roughly seven to ten years, nearly every atom in your body is replaced by a different one.1 The atoms in your brain right now are not the same atoms that were there a decade ago.

And yet you are still you.

This is not a minor detail. This is the single most important fact about your existence, and virtually no one takes it seriously enough.

Think about what this means. The you who is reading this sentence is made of completely different physical stuff than the you who started elementary school. Different carbon atoms. Different oxygen atoms. Different hydrogen atoms. The matter has been swapped out, piece by piece, like replacing every plank of a ship while it's still sailing. And yet at no point during this decades-long replacement process did you ever stop being you. You never flickered out of existence. You never woke up one morning as someone else. The transition was seamless, invisible, and total.

If the atoms themselves are not what makes you you, then what does? This is the question that unlocks everything.

The answer, according to the mainstream view in neuroscience, is **structure and functioning**. You are not your atoms. You are the pattern they form and the processes they carry out. Your consciousness (your thoughts, your feelings, your memories, your sense of being you) arises from the specific way matter is organized and operating in your brain.

The phrase "structure and functioning" is deceptively simple. Structure refers to how the matter in your brain is physically arranged at any given moment: the placement and types of atoms and molecules, the geometry of your neural connections, the concentrations of neurotransmitters at your synapses, even the quantum states of your particles. Functioning refers to how that structure changes over time: the electrical signals firing along your neurons, the chemicals flowing across synapses, the constant dynamic activity that constitutes a living, thinking brain.

Together, structure and functioning encompass everything physical about your brain. And according to everything we know about neuroscience, structure and functioning are what produce your conscious experience.

This is not controversial. It is the mainstream scientific view. But almost no one has followed this insight to its remarkable conclusion.

A note before we continue. Everything up to this point (the atomic replacement, the role of structure and functioning) is empirical observation, well-supported by neuroscience. From here, we move to a proposed interpretation of these observations. The interpretation is logical rather than speculative, but it is an interpretation, and I want to flag the transition rather than blur it.

A pattern can, in principle, be reproduced. A pattern can exist again.

Death destroys the structure of your brain. It does not destroy the possibility of that structure existing.

This is the crack in the foundation of everything you thought you knew about mortality. And this book is going to explore what that crack reveals.

The Two Deaths

It is worth pausing here to make a distinction that most people, and even most philosophers, fail to make. There are two very different things we might mean by "death."

The first is what we might call **external death**: death as seen by other conscious beings. When your body stops working, the people around you observe that you are no longer animated, no longer communicating, no longer behaving like a conscious being. From their perspective, you are gone. This is death as we normally understand it.

The second is what we might call **experiential death**: death from your own conscious perspective. This is the end of your experience. The cessation of your ixperiencitness (a term we will define carefully in Chapter 8). It is the moment after which you never experience anything again.

Most people assume these two kinds of death are the same event. Your body dies, and your experience ends. Simple.

But they are not the same event. And the difference between them is everything.

External death is observable, measurable, and well-understood. A doctor can pronounce you dead. An autopsy can explain why you died. Your body is right there, and it has stopped working. There is no mystery here.

Experiential death is a completely different question. What happens to your experience when your brain stops functioning? The obvious answer ("it stops") is only obvious if you believe that your experience is tied to your specific body. If you accept that your experience is produced by structure and functioning, and that structure and functioning can exist in other bodies, then the "obvious" answer is not obvious at all.

We see each other die. We observe external death constantly. But we never observe experiential death, because by definition, we can only observe our own experience from the inside, and if our experience has truly ended, there is no one there to observe the ending.

This gap between external death and experiential death is where the entire theory lives.

What You Cannot Experience

Here is a fact that sounds trivial but turns out to be profound: **you can never experience not existing.**

Think about it. To experience something, you must be conscious. To be conscious, you must exist (as a conscious being). Therefore, every experience you ever have is, by definition, an experience of existing. You cannot have the experience of non-existence, because there is no "you" there to have it.

This means that from your own perspective (the only perspective that matters for the question of your survival) you will never encounter your own death. You will experience being alive, and then... you will experience being alive. Because that is the only kind of experience there is.

This is not a trick of language. This is a deep fact about the nature of consciousness. And it has implications that most people have never considered.

If you were put into suspended animation (frozen in a state of perfect preservation, with no brain activity, no consciousness) for a billion years, and then revived with your brain structure and functioning perfectly restored, what would that billion years feel like to you?

Nothing. Absolutely nothing. From your perspective, no time would have passed. You would go from one moment of consciousness to the next, with the billion-year gap simply not existing in your experience.

Now ask yourself: Is there any difference, from your experiential perspective, between being frozen for a billion years and then revived, versus simply ceasing to exist for a billion years and then having your brain's structure and functioning reproduced in a completely new body?

If you are the pattern, and not the atoms, the answer is no. There is no difference. In both cases, you experience a gap of nothingness (which you do not experience at all), followed by the resumption of consciousness. The fact that the second scenario involves different atoms is irrelevant. If the pattern is the same, the experience is the same.

This is the key insight. And everything else in this book follows from it.

If you are the pattern and not the atoms, then death destroys one instance of the pattern but not the pattern itself. The question is not whether you end, but what "you" really means — and answering that requires us to look carefully at what science actually knows about consciousness.

CHAPTER 2

The Three Premises — And What Science Says About Them

Before we go further, we need to establish what science actually knows about consciousness, because the entire argument that follows rests on scientific premises, not speculation.

Let me be frank: consciousness is one of the great unsolved problems in science. We do not have a complete theory of how subjective experience arises from physical processes. This is sometimes called "the hard problem of consciousness," and it remains a topic of vigorous scientific and philosophical debate.

But the fact that we don't have a complete theory does not mean we know nothing. In fact, we know quite a lot, and what we know is more than sufficient to support the arguments in this book. Here is what the evidence strongly supports:

Premise One: The Production Premise

The structure and functioning of the brain produces behavior, consciousness, and the quality of subjective experience. When you damage the brain, you alter consciousness. Damage the visual cortex, and you lose the ability to see. Harm the hippocampus, and you lose the ability to form new memories. Injure the prefrontal cortex, and your personality changes. Administer anesthesia, and consciousness disappears entirely. When the anesthesia wears off, consciousness returns. The correlation between brain functioning and conscious experience is so tight and so well-documented that it is, for all practical purposes, a well-established finding.

This evidence comes from multiple independent lines of research:

Brain lesion studies. For over a century, neurologists have documented how specific brain injuries affect specific aspects of consciousness. Phineas Gage, the famous railroad worker who survived an iron rod through his frontal lobe in 1848, became a different person after his injury (same body, same atoms, but different personality, different behavior, different conscious experience). His case, and thousands like it, demonstrate that consciousness is not some ethereal thing floating above the brain. It is produced by specific brain structures doing specific things.

Neuroimaging. Modern brain scanning technologies (fMRI, PET scans, EEG) allow us to watch the brain in real time as it produces conscious experience. We can literally see different regions of the brain light up as you think different thoughts, feel different emotions, or perceive different sensations. The correspondence between brain activity and conscious experience is systematic, reproducible, and difficult to dismiss.

Pharmacology. Drugs alter consciousness by altering brain chemistry. Caffeine makes you alert by blocking adenosine receptors. Alcohol impairs judgment by depressing neural activity in the frontal lobes. Psychedelics produce vivid hallucinations by modulating serotonin receptors. Anesthetics eliminate consciousness entirely by disrupting neural communication across the brain. Every one of these effects is a demonstration that consciousness is produced by, and dependent on, brain functioning.

Anesthesia experiments. These deserve special attention because they are so powerful. When a person is put under general anesthesia, their consciousness is completely eliminated, more thoroughly than in sleep. There is no experience of passing time. There is no "you" during anesthesia. And yet, when the anesthesia wears off and the brain resumes normal functioning, consciousness returns. You are back.

This tells us something important: consciousness is not a permanent property of your body. It is something your brain does when it functions in the right way. Stop the right functioning, and consciousness stops. Restore the right functioning, and consciousness returns. The implications of this will become clear.

Neurosurgery. Perhaps the most dramatic evidence comes from brain surgery performed on conscious patients (a common practice, since the brain has no pain receptors). Surgeons can stimulate specific brain regions with electrical probes and produce specific conscious experiences: vivid memories, emotions, visual percepts, even the sensation of being outside one's own body. This is direct, causal evidence that brain activity produces conscious experience.

The conclusion is strongly supported: consciousness is produced by the structure and functioning of the brain. This is Premise One, and the evidence for it is as robust as for any major claim in neuroscience.

A clarification is important here. None of these results settle the hard problem of consciousness, the question of why physical processes are accompanied by subjective experience at all. What they do is constrain viable theories. Any theory of consciousness must be consistent with the tight, systematic relationship between brain structure/functioning and conscious experience. The evidence does not tell us why brains produce consciousness. It tells us, very convincingly, that they do.

Premise Two: The Multiplicity Premise

The specific atoms don't matter; the structure and functioning do, and many different physical configurations can produce identical conscious experiences. As we discussed in Chapter 1, your atoms are constantly being replaced, yet your consciousness continues. What persists is not the material, but the organization. A neuron firing is a neuron firing, regardless of which specific carbon atoms happen to be in that neuron at the moment.

This principle is not unique to consciousness. It applies to everything in physics and chemistry. Two water molecules composed of different hydrogen and oxygen atoms are nonetheless identical in their properties. Two computer processors manufactured from different batches of silicon will run the same software identically. A recipe yields the same dish whether cooked in one kitchen or another. The specific matter does not contribute some unique, irreplaceable essence; it is the arrangement and operation that determine properties.

The scientific term for this idea is **substrate independence**. A property is substrate independent if it depends on the organization of matter rather than the specific matter itself. Computation is substrate independent: the same calculation can be performed on different computers made of different materials. Information is substrate independent: the same information can be stored on paper, on a hard drive, on a DNA strand, or in a pattern of neural connections.

Consciousness, the evidence strongly suggests, is substrate independent too. It depends on the structure and functioning of the brain, not on which particular atoms happen to make up that brain at any given moment.

What does "structure and functioning" actually include? The answer is: everything physical. The term, as used in Ixperiencit Theory, encompasses:

- How the matter in the brain is arranged at the present moment

- What kinds of elements and molecules are used, and their placement relative to each other
- The excitation states of atoms and other quantum mechanical aspects
- How the neurons are wired together (the connectome)
- The geometry and topology of neural connections
- Individual synaptic properties: their number, size, placement, types of neurotransmitters
- The shapes and branching patterns of individual neurons
- Cellular chemical composition
- How all of these things are changing over time (the dynamic activity that constitutes brain functioning)

In short, structure and functioning includes every physical property of the brain that is relevant to its operation. And it is this totality, not the specific atoms, that produces consciousness.

But there is a further implication worth making explicit. From any given moment of consciousness, an enormous number of different future conscious lifetimes are possible, and many different physical configurations would produce identical subjective experiences. Right now, an enormous number of different things could happen to you in the next second. You could keep reading. You could look up. A meteor could strike nearby. Each possibility leads to a different sequence of experiences. But many slightly different physical scenarios would lead to exactly the same conscious experience from your perspective. Whether the air molecules to your left shift slightly one way or another makes no difference to what you experience. The space of possible futures that produce identical experiences is vast.

This is not a wild claim. It follows directly from the fact that consciousness depends on brain structure and functioning, and that extremely small variations in the body (like the orientation of a single atom in your left knee) have zero effect on what your brain does. There are an enormous number of slightly different physical states that produce the same conscious state. The mapping from physics to consciousness is many-to-one, not one-to-one.

Premise Three: The Identity Premise

Identical structure and functioning produce identical properties. This is a foundational principle of physics and chemistry. Two water molecules with different atoms but identical structure have identical properties. Two identical computer programs running on identical hardware produce identical outputs. Applied to any physical property other than consciousness, this principle is uncontroversial. The question is whether consciousness follows the same rule — and the evidence strongly suggests it does.

Applied to consciousness, this premise says: **identical brain structure and functioning, regardless of where, when, or in what matter it occurs, would produce identical consciousness.**

Some people find this premise obvious. Of course identical physical systems produce identical results — that's what "identical" means. Others find it deeply unsettling when applied to consciousness because it implies that their subjective experience is not uniquely tied to their particular body.

Both reactions are correct. It is obvious, from a physics perspective. And it is unsettling, from a personal perspective. But unsettling is not the same as wrong.

I should note that many philosophers reject Premise Three when applied to consciousness specifically, even while accepting it for every other physical property. They argue that consciousness might involve something beyond structure and functioning, something that makes each instance unique in a way that, say, two water molecules are not. We will engage with the strongest version of this objection in Part Four. For now, I ask only that you hold the premise provisionally and see where it leads.

An identical brain, built from completely different atoms, in a completely different location, at a completely different time, functioning in the same way as yours, would produce the same conscious experience. It would *be* you, in every way that matters. Not a copy that merely resembles you. Not an imitation. The same conscious experience, because consciousness *is* the structure and functioning, and the structure and functioning are the same.

This claim involves both physics (the premises about brain structure) and philosophy (the inference about identity). But the philosophical step is modest: it simply applies to consciousness the same principle we already accept for every other physical property.

The phrase "sufficiently similar" is important here. Perfect atomic-level identity is almost certainly not required for identity of conscious experience, just as you don't need every grain of sand to be in the same position for two identical sandcastles to have the same shape. There is a tolerance, a margin of physical variation within which the conscious experience remains the same. Determining the exact boundaries of this tolerance is a scientific question that future research will need to address, but the principle is clear.

What These Premises Imply

These three premises are individually mainstream. The Production Premise is accepted by virtually every working neuroscientist. Applied to any other physical property, the Identity Premise is accepted by every physicist. And the Multiplicity Premise is a logical observation about the space of possibilities.

But together, they lead to a conclusion that is anything but mainstream:

You are not limited to one body, one lifetime, or one moment in time. Your consciousness, your ixperiencitness, can in principle be produced by any brain structured and functioning

like yours, wherever and whenever the right physical conditions are met.

I want to be transparent about something. The three premises are third-person claims about what happens when brains with certain structures exist. The conclusion about your immortality is a first-person claim about what you will experience. The bridge between them is the identity thesis: that if a future brain produces the same ixperiencitness as yours, the being having that experience is you, in the only sense that matters. This bridge is not a hidden assumption; it is a direct consequence of the premises, if you take them seriously. But it is the place where many readers will feel the strongest resistance. We will return to it from different angles throughout the book.

It means that you are, in a rigorous and well-defined sense, immortal. The chapters that follow will show you why the logic is sound and the conclusion is difficult to escape. But first, we need to understand why this conclusion is so hard to accept, and why the difficulty of accepting it is not evidence against it.

What About the "Hard Problem"?

At this point, a philosophically inclined reader might object: "But what about the hard problem of consciousness? We don't understand how brain structure and functioning produce subjective experience. Maybe consciousness involves something beyond the physical. Maybe there's an extra ingredient, something like a soul, even if we don't call it that."

This is a fair point, and I want to address it directly.

Yes, we do not have a complete scientific explanation of how subjective experience arises from physical processes. This is a genuine gap in our understanding. But here is what matters for our argument: we don't need a complete explanation. We only need the three premises stated above, and the evidence for those premises is overwhelming regardless of whether we can fully explain the mechanism.

Consider an analogy. In the 18th century, scientists did not understand how gravity worked. They knew that massive objects attracted each other, and they could predict the strength of this attraction with great precision, but they had no idea why mass produced gravitational attraction. Newton himself said he "feigned no hypotheses" about the underlying mechanism.

Did this lack of mechanistic understanding invalidate the law of gravity? Of course not. The evidence that gravity existed, and that it behaved in predictable ways, was overwhelming. The fact that scientists couldn't explain why it worked didn't change the fact that it did work.

The same applies to consciousness. We may not understand exactly how structure and functioning produce subjective experience. But we have extensive evidence that they do, and that identical structure and functioning produces identical consciousness. That evidence is sufficient for everything that follows in this book.

Moreover, any proposed "extra ingredient" beyond structure and functioning faces a devastating problem: there is no evidence for it. Every piece of evidence we have (from brain lesions, from neuroimaging, from pharmacology, from anesthesia, from neurosurgery) points to structure and functioning as the complete explanation. Proposing an extra ingredient is exactly like proposing a soul. It might be comforting, but it has no scientific basis.

Each premise is individually mainstream. Together they imply something that is not: that your consciousness is not confined to one body, one location, or one moment in time. But before we follow these premises forward, we need to reckon with why the conclusion feels so wrong.

CHAPTER 3

Why Your Intuitions Are Wrong

There is a powerful argument against everything I have just said, and it goes like this: "That can't be right. It just doesn't feel right. I know I am this body, this brain, these atoms. A copy of me would not be me, no matter how perfect. It's obvious."

This is what philosophers call the argument from intuition. And it is perhaps the most common reason people reject the ideas in this book without actually engaging with the logic.

So let me explain why your intuitions about consciousness and identity are almost certainly wrong, and why trusting them on this topic is about as wise as trusting your gut feeling that the Earth is flat.

The Copernican Lesson

There was a time when the most obvious fact in the world was that the sun moved across the sky. You could see it with your own eyes. It rose in the east, traveled in an arc overhead, and set in the west.

The Earth was clearly stationary (you couldn't feel it moving) and the sun clearly orbited around it.

This "obvious" fact was wrong. The Earth rotates on its axis and orbits the sun. But the truth was so contrary to everyday experience that it took centuries of scientific investigation, mathematical proof, and bitter controversy before it was widely accepted.

The lesson of the Copernican Revolution extends far beyond astronomy. It is about the reliability of human intuition when it comes to questions that go beyond everyday experience.

I want to be clear: intuition is not useless. It is extraordinarily valuable within its evolutionary domain. Your intuitions about social dynamics, physical danger, and everyday cause-and-effect are remarkably accurate, because those are the domains evolution shaped them for. But our intuitions about identity and consciousness are operating outside that domain. They were shaped by evolution to help us survive and reproduce in a specific environment — the African savanna, thousands of years ago. They were not shaped to help us understand the deep nature of consciousness, any more than they were shaped to help us understand planetary motion.

Evolution gave us a strong sense of being a single, continuous, unique self, housed in a single body. This sense was useful for survival. An organism that treated its own body as special, that fought to defend it, fed it, reproduced with it, had a survival advantage over an organism that didn't. Natural selection built this sense of singular selfhood deep into our psychology.

But useful for survival is not the same as true. The sense that the sun orbits the Earth was useful too; it helped our ancestors predict the seasons and navigate by the stars. That didn't make it true.

The Folk Psychology Trap

Psychologists have a term for the set of intuitive beliefs we all have about how the mind works: **folk psychology**. Folk psychology includes beliefs like "I am a single self," "my consciousness is continuous," "I am tied to this specific body," and "a copy of me would not really be me."

These beliefs feel self-evident. They feel like they must be true because of how obvious they seem. But this feeling of obviousness is exactly what should make us suspicious.

Consider the following folk-psychology beliefs that turned out to be wrong:

- **"We see the world as it really is."** Wrong. Our visual system constructs a model of the world based on limited sensory data, filling in gaps, correcting for distortions, and even fabricating details. Visual illusions prove this conclusively. The world you see is not the world as it is; it is a construct of your brain.
- **"Our memories are like recordings."** Wrong. Memories are reconstructed every time we recall them, and they are highly susceptible to distortion, fabrication, and suggestion. Eyewitness testimony is one of the least reliable forms of evidence, despite feeling utterly certain to the person remembering.
- **"We make decisions based on rational deliberation."** Wrong. An enormous body of research in behavioral economics and cognitive psychology shows that most of our decisions are made by unconscious processes, with conscious reasoning serving mainly to rationalize decisions already made.

In each case, the folk-psychology belief felt obviously true. And in each case, it was wrong. Why should we expect our folk-psychology beliefs about identity and consciousness to be any different?

The "It's Not Me" Argument

The most common folk-psychology objection to the ideas in this book is what I call the "It's Not Me" argument. It goes like this:

"Even if you built a perfect replica of my brain, made of different atoms, in a different place, it would not be me. It would be a copy. I would still be here, in this body. The copy might think it's me, but it would be wrong."

This argument feels compelling. But let's examine what it actually claims.

The "It's Not Me" argument asserts that there is something about your specific body, something beyond its structure and functioning, that makes your consciousness yours. It asserts that even if two brains are physically identical in every measurable way, they could produce different conscious experiences, because one of them is the "original" and the other is a "copy."

But what could this "something" be? It's not the atoms, because your atoms are constantly being replaced. Nor is it the structure, because the structure is identical by hypothesis. And it's not the functioning, because the functioning is identical by hypothesis.

The only thing left is the history of the body: the fact that one body has been continuously existing since birth, while the other was just created. The "It's Not Me" argument, at bottom, is an argument that the history of a body matters more than its present physical state.

But does the history of a body affect its present consciousness? Consider this: if you learned that, without your knowledge, every atom in your body had been replaced simultaneously last night while you slept, replaced with identical atoms in the same positions — would you be a different person this morning? Would you notice? Would anything about your conscious experience be different?

The answer, obviously, is no. Your consciousness is determined by the present structure and functioning of your brain, not by the history of the atoms that make it up. History matters only insofar as it shapes the present physical state. If the present physical state is the same, the consciousness is the same, regardless of how that state was achieved.

This is like saying that the same recipe, followed precisely in two different kitchens with different brands of flour and different ovens, produces two "different dishes," even though the ingredients, proportions, and results are identical. We can have identity of properties — identity of consciousness, identity of experience — without identity of the physical objects that produce those properties.

The "It's Not Me" argument, when you strip away the intuitive appeal, is really an argument for a soul. In its simplest form, it posits something beyond the physical, some undetectable essence that makes this body special and no other body acceptable. Science has a word for such a non-physical, untestable essence: the supernatural. And there is no evidence for it.

But there is a more rigorous version of this objection, developed by the philosopher Derek Parfit, that does not require anything supernatural. It argues that what matters is not physical identity but causal continuity, the unbroken chain of cause and effect that connects one moment of your consciousness to the next. The causal-continuity requirement sounds compelling, but consider: your body already replaces its atoms one at a time through normal metabolism, and at no point does the causal chain break — yet the end result is a completely different set of atoms producing your consciousness. Causal continuity is preserved even as the material is entirely swapped out. This suggests that causal continuity, properly understood, tracks structure and functioning rather than specific matter — which is exactly what Ixperiencit Theory claims. (We examine Parfit's argument in full depth in Chapter 18, but the short answer is that causal continuity, when you follow its logic rigorously, supports our conclusion rather than undermining it.)

How to Think Clearly About Identity

So if our intuitions are unreliable, how should we think about identity?

The answer is: by following the evidence, even when it leads somewhere uncomfortable.

Here is a simple method. When you find yourself thinking "but that's not really me," ask yourself: What experiment could I perform to test this claim? What observable, measurable, physical property would distinguish the "real me" from the "not really me"?

If you cannot identify any such property (if the two systems are physically identical in every detectable way, produce identical behavior, give identical answers to every question, and report identical subjective experiences) then the claim that one is "really you" and the other is "not really you" is scientifically empty. It is like arguing about whether the number 2 written on one piece of paper is the "real" 2, while the number 2 written on another piece of paper is just a "copy." The question is meaningless, because the property (the number 2, the conscious experience) is not tied to the specific physical substrate.

This principle, that claims about identity must be grounded in observable, testable properties, is what separates scientific thinking from folk psychology. And once you adopt it, the conclusions of this book become difficult to avoid.

In the next part of the book, we will build the theoretical framework that makes these conclusions precise. We will define key concepts, explore the structure of conscious experience, and develop a vocabulary for talking about consciousness with the clarity and precision that the topic demands.

Your intuitions about identity were built for the savanna, not for the truth. Once you let go of them, the path from the three premises to their conclusion becomes very difficult to block.

PART TWO:
THE THEORY

CHAPTER 4

You Are Not Your Atoms

To understand why you can never truly die, you first need to fully absorb a simple but important truth: **the matter itself does not perpetuate the self. It is the structure and functioning of matter that perpetuates the self.**

But before we dig into the evidence, it helps to know where this claim fits in the philosophical tradition. Philosophers have proposed four major theories of personal identity:

The Soul Theory holds that you are an immaterial soul attached to your body. Identity persists because the soul persists. This view is intuitive and comforting, but it has no scientific support. There is no evidence for any non-physical substance that carries identity.

The Body Theory holds that you are your physical body. Identity persists as long as the same body persists. This view has the advantage of simplicity, but it cannot explain why you remain "you" after every atom in your body has been replaced.

The Psychological Continuity Theory, developed most rigorously by Derek Parfit, holds that you are a chain of psychological connections: memories, personality traits, intentions, beliefs. Identity persists as long as these connections persist, even across changes in the body. This is closer to the truth, but it faces the question: what produces these psychological connections? Parfit stopped at the psychology. Ixperiencit Theory goes further.

The Structure-and-Functioning Theory, the view of this book, holds that you are the physical pattern of your brain: its structure and its dynamic activity. Psychological connections persist because brain structure persists. Consciousness is produced because matter is organized in the right way. This view inherits the strengths of the psychological continuity theory while grounding it in something physical, measurable, and — importantly — reproducible.

With this map in hand, let us build the theory.

Consider an analogy. A song is not the CD it is recorded on. It is not the vinyl grooves of an LP, not the magnetic patterns on a cassette tape, and not the digital file on your phone. The song is the pattern, the specific sequence of frequencies and rhythms. You can destroy the CD and the song still exists, because the pattern can be reproduced in a different medium.

Your consciousness is like this, but far more so.

The atoms in your body are borrowed. You borrow carbon from the food you eat, oxygen from the air you breathe, water from what you drink. Your body uses these atoms for a while, arranged in very specific structures that carry out very specific functions, and then releases them back into the world. New atoms come in and take their place, slotting into the same structures, carrying out the same functions.

At no point during this constant atomic turnover do you stop being you. The pattern persists even as the material changes.

The Chemistry of Replacement

Let's put some numbers on this. Your body takes in matter in the form of air, liquids, and food, and expels it in the form of exhaled breath, urine, sweat, feces, and shed skin cells. Chemistry is constantly assimilating new matter into the structure of the body, replacing old matter that is being expelled.

But not all matter is replaced at the same rate. The water in your blood can be replaced in hours. The cells lining your gut are replaced roughly every five days. Your skin cells turn over every two to three weeks. Red blood cells last about four months. The cells in your liver are replaced roughly once a year. Even the seemingly permanent structure of your bones is continuously remodeled, with the mineral content being fully replaced over the course of about ten years.2

What about the brain? Neurons are among the longest-lived cells in the body, and most neurons do persist for a lifetime without being replaced. But (and this is the key point) the atoms within those neurons are replaced. The proteins, lipids, and other molecules that make up each neuron are in constant flux, turning over on timescales of days to weeks. Neurotransmitter molecules at your synapses last only milliseconds before being recycled. And the water molecules in and around your neurons are replaced on timescales of seconds to minutes.

So while the cells in your brain may persist, the matter within those cells does not. The atoms in your neurons today are not the atoms that were there a year ago. And yet your consciousness persists, your memories persist, your sense of self persists. Because the pattern persisted even as every atom changed.

The Speed of Replacement: A Thought Experiment

How fast can matter be replaced in a body and still maintain the same person?

Philosophers have constructed thought experiments about this question, and they are revealing. At one extreme, we have normal biological replacement, the slow, continuous turnover that happens naturally. Everyone agrees that this preserves identity. You are the same person despite the gradual replacement of your atoms.

At the other extreme, imagine instantaneous replacement: every atom in your body is simultaneously swapped out for an identical atom of the same element, in the same position, in the same quantum state. The body before and after the swap would be physically indistinguishable. Would you still be you?

The answer must be yes. The two states are physically identical. There is no measurable difference. Any claim that you would be a different person after the swap is equivalent to claiming that the specific atoms themselves, apart from their type, position, and state, carry some kind of identity. And there is no physical basis for this. Atoms of the same element in the same state are, according to quantum mechanics, truly identical. They are not merely very similar; they are literally indistinguishable. This is a foundational principle of quantum physics called the indistinguishability of identical particles — a principle formalized in the spin-statistics theorem and confirmed by every quantum experiment ever conducted.

To be precise: quantum indistinguishability means there is no physical experiment that could distinguish one hydrogen atom from another in the same state. Whether this constitutes "metaphysical" identity is a philosophical question, but for our purposes, what matters is the physical fact: no experiment can tell the difference.

So the speed of matter replacement is irrelevant. Whether it takes ten years (as in normal biology) or ten nanoseconds (as in our thought experiment), the result is the same: you are the same person, because the structure and functioning are the same.

This conclusion has a staggering implication: **the specific atoms in your body right now play no role in making you *you*. They could all be replaced simultaneously, and you would be unchanged. What matters is the pattern, and only the pattern.**

The Song and the iPhone

Imagine you are listening to a beautiful song on your iPhone. You love this song. It moves you deeply. Now imagine that your iPhone breaks, and you buy a new one. You download the same song on the new iPhone and play it. Is it the same song?

Of course it is. Different physical device, different silicon chips, different glass screen, different aluminum case. But the song is the same, because the song is not the iPhone. The song is the pattern of sound waves that the iPhone produces. Any iPhone that produces the same pattern produces the same song.

This analogy, which was one of the first I used to explain these ideas, is illuminating, but it's important to understand where it works and where it doesn't.

Here's where it works perfectly: just as the same song can be produced by different iPhones, the same consciousness can be produced by different brains. Just as the destruction of one iPhone doesn't destroy the song (because other devices can play it), the destruction of your brain doesn't destroy your consciousness (because other brains could produce it). Just as the iPhone can be enhanced to produce better sound quality, brains could be enhanced to produce richer conscious experience.

Here's where the analogy needs to be extended: a song is relatively simple (it's a fixed sequence of sound waves). Consciousness is vastly more complex: it is a dynamic, ever-changing process involving trillions of interactions. And the key thing we care about with consciousness is not just whether it sounds the same to an outside observer, but whether it feels the same from the inside. This is the question of ixperiencitness, which we will explore in Chapter 8.

But the fundamental principle holds: **you are to your brain as a song is to an iPhone. You are the pattern, not the device. And patterns don't die when devices break.**

The atoms cycle through your body on timescales from seconds to years; quantum mechanics confirms they are literally interchangeable; and whether they are swapped gradually or all at once, nothing about your experience changes. But if you are a pattern, then the question of whether a pattern can exist in more than one place, and what that would mean for identity, becomes urgent. The Ship of Theseus, that ancient puzzle about objects and their parts, turns out to be a puzzle about you.

CHAPTER 5

The Ship of Theseus — Applied to You

In ancient Athens, there was said to be a ship that had belonged to the hero Theseus.[3] The Athenians preserved it in a harbor as a monument, and as the years passed, the salt air and sea did their work. When a plank rotted, the Athenians replaced it with fresh timber. Then another plank. Then the mast. Then the keel. Over the centuries, every single piece of the original ship was replaced. Historians, tourists, and philosophers stood on the dock and debated: **Is it still the same ship?**

Now imagine a collector who, unbeknownst to the Athenians, saved every rotted plank as it was removed. In a warehouse across the harbor, the collector slowly reassembled the original timber into a complete ship — warped, salt-stained, and held together with care. Eventually the collector has a ship made entirely of Theseus's original material. And in the harbor sits a ship with Theseus's exact form and function, made of entirely new wood.

Which is the Ship of Theseus? The one that has the original form but new matter? Or the one that has the original matter but a reassembled form?

This question, known as the Ship of Theseus paradox, has puzzled philosophers for over two thousand years. But when we apply it to human bodies and consciousness, it becomes far more than an academic puzzle. It becomes the key to understanding immortality.

The Basic Paradox

Here is the Ship of Theseus, applied to you.

Your body is constantly replacing its own matter. Over the course of roughly a decade, nearly every atom in your body has been swapped out. You are like the ship in the harbor: every plank has been replaced, but you are still "you."

Most people are comfortable with this. The replacement is gradual, and there is continuity of the overall structure. It feels like the same ship, even though no original timber remains.

But now let's make it personal.

The Matter-Swap Experiment

Imagine the following scenario. You are the **original**, the person whose identity we are tracking. Somewhere else, there exists a **continuing replica** of you, a body that not only matches your physical organization at one moment but continues to match it over time. This replica has exactly the same structure and functioning as you, down to every atom. It was built from completely different matter, but it is physically identical to you in every measurable way.

Picture two identical bodies lying on tables in a laboratory, side by side. Same height, same weight, same face. If you could peer inside their brains with a sufficiently powerful microscope, you would see the same neurons, the same connections, the same chemical balances. One is you. The other was built from scratch.

Now imagine that we slowly exchange parts between you and your continuing replica. One atom at a time, we take an atom from your body and replace it with the corresponding atom from the continuing replica's body, and vice versa. We do this simultaneously, keeping both bodies functioning identically throughout the process.

After the exchange is complete, your body now contains all the matter that was originally in the continuing replica, and the continuing replica's body contains all the matter that was originally in you.

Which one is now the original?

If identity is determined by **bodily continuity** (the continuous existence of one particular body over time) then the answer is straightforward: you are still the original, because your body is the one that has been continuously existing in the same location. The fact that your matter has been completely replaced doesn't change your identity, because (as we've established) your atoms aren't what make you you.

But if identity is determined by **matter** (by the specific stuff you're made of) then the answer is reversed: the continuing replica is now the original, because it now contains all of your original atoms.

These two criteria give contradictory answers. Which one is right?

The answer, according to the science of structure and functioning, is: neither criterion is what actually matters. What matters is the **physical organization and activity** of the body, particularly the brain. And since both bodies have had identical organization and activity throughout the process (we specified this as a condition of the experiment), **both bodies produce the same consciousness, and both have equal claim to being "you."**

This is the central lesson of the Ship of Theseus, applied to consciousness: identity is not determined by matter, and it is not determined by bodily continuity. It is determined by the pattern. And the pattern can be present in more than one body at a time.

The Split-and-Recombine Experiment

Now consider an even more provocative version of the experiment. Instead of exchanging atoms one at a time, imagine splitting both bodies down the middle (left half and right half) and recombining them: the left half of the original with the right half of the continuing replica, and vice versa.

Which of the two resulting bodies is the original?

Neither has full bodily continuity with the original (each has only half). Neither contains all of the original matter (each contains only half). There is no principled way to say that one is "you" and the other is "not you," based on either matter or bodily continuity.

But if what matters is the pattern, and if both resulting bodies can be made to have the same organization as the original (by carefully constructing the reassembly), then both can produce the same consciousness.

These thought experiments are not meant to be practical proposals. We are nowhere near being able to swap atoms between bodies or split bodies in half. They are meant to reveal the logical implications of our premises, by pushing them to extremes. And what they reveal is that the common-sense notion of identity (one body, one self, end of story) is not just wrong. It is incoherent. It breaks down under scrutiny.

The coherent alternative is the one this book proposes: **you are the pattern, not the body. You are the organization, not the material. And the pattern can exist in more than one place.**

For Death to Be Final

The Ship of Theseus experiments also reveal something important about death.

For death to be truly final, for you to be permanently, irrevocably, forever gone, there must be something about you that can be destroyed permanently and can never be restored or reproduced. Something that is uniquely tied to you and only you.

What could this something be?

It's not the atoms, because atoms are fungible. Nor is it the body as a continuous physical object, because the body's matter is constantly being replaced. It's not your memories, because memories are stored in neural structure, and neural structure can in principle be duplicated. And it's not your personality, because personality is a product of brain functioning, and brain functioning is determined by structure.

The only candidate is some kind of non-physical essence: a soul, a spirit, an irreducible "self" that goes beyond all physical properties. And there is no scientific evidence for any such thing.

If there is no non-physical essence, then everything about you is captured by your physical organization and activity. And if everything about you is captured by that organization and activity, then you can be recreated whenever the pattern is recreated. And if you can be recreated, then death is not final.

The Ship of Theseus, once you apply it honestly to human bodies, demolishes the intuition that identity lives in matter or in bodily continuity. What the thought experiments reveal is starker: for death to be truly final, there would have to be some unreproducible, non-physical essence, and nothing in science supports the existence of such a thing. Everything about you is captured by the pattern, and the pattern can, in principle, be recreated. That is the theoretical claim. But is there real-world evidence for it — evidence from actual brains, not just thought experiments? There is, and it is stranger than any philosopher could have invented.

CHAPTER 6

Your Brain Is Already Proving the Theory

The theory in this book might sound abstract, philosophy dressed up in scientific language. But the most powerful evidence for it comes not from thought experiments but from real cases. Real brains. Real people. Real science. And the cases are far stranger than any thought experiment I could invent.

Phineas Gage: The Man Who Became Someone Else

On September 13, 1848, a 25-year-old railroad construction foreman named Phineas Gage was packing explosive powder into a hole in rock when the charge detonated prematurely. A three-foot-seven-inch iron rod, over an inch in diameter, shot upward through his left cheek, through his brain's left frontal lobe, and out through the top of his skull.4

Gage survived. He was conscious within minutes of the accident, could speak, could walk. But something fundamental had changed.

Before the accident, Gage was described as capable, efficient, well-balanced, and shrewd. After the accident, he was fitful, irreverent, profane, and impatient. His friends said he was "no longer Gage." His physician, John Harlow, noted that "the equilibrium or balance between his intellectual faculties and animal propensities seems to have been destroyed."

Same body. Same atoms (mostly). Different person, or at least, a dramatically different ixperiencitness.

The Gage case demonstrates, in the starkest possible terms, that you are your brain's structure and functioning. Change the structure (destroy part of the frontal lobe), and you change the person. The specific atoms in Gage's frontal lobe were not what made him "Gage"; it was their organization, their connectivity, their role in the larger system of his brain. When that organization was destroyed, the person it produced was destroyed too, even though the body continued to live.

Split-Brain Patients: Two Consciousnesses in One Skull

In the 1960s, neurosurgeon Roger Sperry and his student Michael Gazzaniga began studying patients5 who had undergone a radical surgery for severe epilepsy: the complete severing of the corpus callosum, the thick band of nerve fibers connecting the brain's two hemispheres.

What they discovered was astonishing. After the surgery, each hemisphere seemed to function as an independent conscious entity.

In a classic experiment, a split-brain patient would be shown a picture — say, the word "WALK" — visible only to the right hemisphere (through the left visual field). The left hemisphere, which controls speech, would report seeing nothing. But the patient would stand up and begin walking. When asked why, the left hemisphere would confabulate an explanation — "I needed to get a drink" — unaware that it was the right hemisphere, which had seen the word "WALK," that initiated the action.

In another experiment, the word "LAUGH" was flashed to the right hemisphere. The patient laughed. When asked why, the left hemisphere (unaware of the stimulus) said, "You guys come up and test us every month. What a way to make a living!" The left hemisphere invented a plausible reason for the laughter, completely unaware of the real cause.

What do split-brain cases tell us about ixperiencitness?

They tell us that a single body can house what appears to be two separate streams of conscious experience, two awarepaths running in parallel within the same skull. Each hemisphere has its own perceptions, its own desires, its own intentions. They can even disagree: in some cases, one hand (controlled by one hemisphere) would pull up the patient's pants while the other hand (controlled by the other hemisphere) would pull them down.

This is conscious multiplicity within a single body. And if two consciousnesses can coexist in one skull, the idea that one consciousness can exist in two skulls is not so radical after all.

The Connectome: Mapping the Wiring of You

The Human Connectome Project, launched in 20096, aims to map the complete wiring diagram of the human brain: every neural connection, every pathway, every circuit. It is the most ambitious neuroscience project ever attempted, and its results are directly relevant to Ixperiencit Theory.

What the Human Connectome Project has revealed, through high-resolution diffusion imaging of over 1,200 participants, is that the brain's wiring diagram is both remarkably consistent across individuals (certain pathways and structures are universal) and uniquely individual (the precise pattern of connections differs from person to person, like a fingerprint). Your connectome, the specific way your neurons are wired together, is a critical component of your ixperiencitness.

But the connectome is not the whole story. The same wiring can produce different consciousness depending on:

- **Synaptic strengths:** How strongly each connection transmits signals. These change with experience (learning) and can differ between genetically identical twins.
- **Neurotransmitter levels:** The chemical balance in your brain profoundly affects mood, cognition, and experience. Depression, for example, is associated with altered serotonin signaling: same wiring, different chemistry, different consciousness.
- **Ongoing dynamics:** The brain is not a static circuit. It is a dynamic system in constant motion. The same wiring, with the same chemistry, can produce different experiences depending on what patterns of activity are currently flowing through it.

This is why Ixperiencit Theory emphasizes the complete pattern rather than just structure. The connectome is the hardware; the functioning is the software. Both together produce the consciousness.

Neuroplasticity: Your Brain Rewrites Itself

One of the most important discoveries in modern neuroscience, established through decades of work by researchers like Michael Merzenich and Eric Kandel, is that the brain is far more plastic, more changeable, than anyone previously believed. Your brain is not a fixed machine that you are born with and die with. It is a constantly rewriting, constantly restructuring system.

When you learn something new, your brain physically changes. New synaptic connections form. Existing connections strengthen or weaken. In some cases, entirely new neurons grow (neurogenesis, a phenomenon first demonstrated in adult humans by Peter Eriksson and colleagues in 1998, though the extent remains debated). The brain of a concert pianist is physically different from the brain of a non-musician, different in its actual physical structure, not only in what it knows.7

What does this mean for ixperiencitness? It means that your ixperiencitness is constantly changing. The you of today has a slightly different brain structure, and therefore slightly different ixperiencitness, than the you of yesterday. These changes are usually small enough that you experience continuity of identity. But over decades, the cumulative changes are substantial.

This is directly relevant to the theory. If your ixperiencitness can change while remaining recognizably "you," if there is a continuum of identity rather than a binary, then the reproduction of your consciousness does not require perfect atomic fidelity. It requires fidelity within the range that preserves ixperiencitness. And the range, as neuroplasticity shows, is quite broad.

Anesthesia Revisited: What Actually Happens

We will examine the philosophical implications of anesthesia in a later chapter, but let me first present the neuroscience detail that makes the case so powerful.

Under general anesthesia, the brain does not simply "turn off." Different anesthetic agents work through different mechanisms, but they share a common effect: they disrupt the large-scale integration of information across the brain.8 Individual brain regions may continue to process information locally, but the coordination between regions — the long-range communication that seems to be essential for consciousness — breaks down.

This is significant because it suggests that consciousness depends on global integration, not local neural activity alone. The brain is a system in which specialized modules communicate, and it is this communication — this integration — that produces the unified conscious experience.

Integrated Information Theory (IIT), proposed by neuroscientist Giulio Tononi, formalizes this idea.9 IIT proposes that consciousness is identical to a specific kind of information — "integrated information," or phi — that exists when a system is both highly differentiated (its parts can be in many different states) and highly integrated (its parts are tightly connected). By this measure, a brain under anesthesia has low phi (because integration is disrupted), while a normal waking brain has high phi.

Whether IIT is the correct theory of consciousness remains an open question. But its core insight — that consciousness depends on global integration rather than local computation alone — is supported by a growing body of evidence, including Massimini's TMS-EEG studies showing that integrated cortical processing collapses under anesthesia and in vegetative states. And this insight strengthens the case for Ixperiencit Theory: if consciousness depends on the pattern of integration across the brain, then reproducing that pattern (in any physical substrate) should reproduce the consciousness.

Patient H.M.: Memory Without Identity

In 1953, a patient known as H.M. (later identified as Henry Molaison) underwent surgery10 to treat severe epilepsy. The surgeon removed large portions of both hippocampi — structures deep in the brain essential for forming new long-term memories.

The surgery cured H.M.'s epilepsy. It also destroyed his ability to form new memories. For the remaining 55 years of his life, H.M. lived in a perpetual present. He could remember his childhood and events before the surgery, but every new experience vanished within minutes. He would meet the same doctors every day and greet them as strangers.

H.M.'s case is crucial for understanding ixperiencitness. His personality, his cognitive style, his way of engaging with the world — the core of his ixperiencitness — remained intact. He was still recognizably "himself" to those who knew him. But his awarepath was profoundly altered: instead of accumulating new memories that enriched and changed his conscious life over time, each day was, in an experiential sense, the same day.

This demonstrates that ixperiencitness is not the same as memory. You can lose the ability to form new memories and still retain your ixperiencitness — your characteristic way of experiencing the world. The deep patterns of cognition and personality that make you you are stored differently from episodic memories, in structures that H.M.'s surgery left intact.

What the Evidence Adds Up To

Each of these cases (Phineas Gage, split-brain patients, the connectome project, neuroplasticity research, anesthesia studies, H.M.) tells us something specific about the relationship between brain and consciousness. Taken together, they paint a consistent picture:

1. **Consciousness is produced by brain structure and functioning.** Damage the structure (Gage, H.M.), disrupt the functioning (anesthesia), and consciousness changes or disappears.
2. **Ixperiencitness depends on specific patterns.** Not on specific atoms, but on the way neurons are connected, the way chemicals are balanced, and the way information flows. Change the pattern enough, and you change the person.
3. **Identity is a continuum, not a binary.** Gage after the accident was not completely different from Gage before — he retained some memories, some abilities, some recognizable traits. But he was significantly different. Identity is a matter of degree, not all-or-nothing.

4. **The brain is far more flexible than our intuitions suggest.** It can house two consciousnesses (split-brain). It can change its structure through experience (neuroplasticity). It can function without memory formation (H.M.). The folk-psychology picture of one fixed self in one fixed brain is far too simple.
5. **Global integration matters.** Consciousness seems to depend on how brain regions communicate, not on what individual regions do in isolation. This supports the idea that consciousness is a property of the system, not of any individual component — and that any system with the right integration could produce it.

This is not abstract philosophy. Every case — Gage, the split-brain patients, H.M., the plasticity research — points the same way: change the pattern and you change the person; reproduce the pattern and you reproduce the consciousness. With that evidence in hand, we are ready to define the concepts we need to talk about consciousness, identity, and survival.

CHAPTER 7

The Awarepath: Your Conscious Lifetime

Imagine you could rewind the movie of your life. Not the external events, not the home videos or the photographs, but the inner movie. Every sight, every sound, every taste, every flash of emotion, every half-formed thought, every dream you forgot by morning, all of it, from the very first flicker of awareness you ever had to the last moment of consciousness you will ever experience. Played back in order, in full, with nothing left out.

That is your **awarepath**.

What Is an Awarepath?

To get a feel for what an awarepath actually contains, slow the movie down. Pick a single moment, say, right now. What is in your awareness at this very instant?

There is the visual field: these words on a page or screen, the peripheral blur of whatever surrounds them, the color of the light in the room. Sound reaches you: maybe traffic outside, maybe music, maybe the hum of silence itself. You feel your body — the pressure of a seat against your back, the weight of your arms, the temperature of the air on your skin. A taste lingers in your mouth, or the absence of one. A smell you have probably stopped noticing hangs in the air.

And underneath all of that, there is the cognitive layer: the thread of meaning you are following in this sentence, a background awareness of what time it is, perhaps a flicker of anxiety about something you need to do later, or a low hum of contentment, or boredom, or curiosity. All of it exists simultaneously, layered on top of itself, woven together into a single moment of experience.

Now multiply that by every waking second of your life. Every moment of childhood wonder and teenage angst and adult routine. Every dream. Every daydream. The dull moments and the vivid ones, the ones you remember and the countless millions you have forgotten. Strung together, in the order you lived them, they form one continuous thread — your awarepath.

Your awarepath is the complete sequence of all your conscious moments, from the first flicker of consciousness you ever experienced to the last. It is everything you have ever thought, felt, perceived, dreamed, or imagined, arranged in the order you experienced it. It is the totality of your conscious life.

Right now, at this very moment, you are at one point along that thread. Behind you lie all the conscious moments you have already experienced. Ahead of you lie all the conscious moments you have yet to experience.

But here is what makes the awarepath concept so powerful: **your awarepath is not necessarily limited to what one body produces.**

In standard thinking, your conscious life starts when your brain first produces consciousness (sometime before birth or during infancy) and ends when your brain stops functioning (at death). Your awarepath, in this view, is simply the conscious experiences produced by one brain over one lifetime.

But Ixperiencit Theory shows that this view is far too narrow. If identical brain structure and functioning produces identical consciousness, then your awarepath potentially extends to every instance in the universe where a brain structured and functioning like yours produces conscious experience. Your conscious life is not one line; it is a vast branching tree, with each branch representing a different physical body that produces the same conscious experience you would have.

The Layers Beneath: Physical Path and Neural Path

The awarepath, your inner movie, does not appear out of nowhere. It is produced by a physical system: your brain, housed in your body. To understand how conscious experience relates to the physical world, it helps to name these underlying layers.

Your body traces its own path through time. From conception to death, every atom in your body occupies a specific position at each moment, every chemical reaction proceeds along a particular course, every cell divides or dies on a particular schedule. This complete physical journey — the trajectory of your entire body through the space of possible physical states — is your **physical path** (formally, physapath in the full theoretical framework).

Nested within the physical path is a more specific trajectory: that of your brain. The organization and activity of your brain over time — every neural connection, every firing pattern, every surge of neurotransmitter, constitutes your **neural path** (formally, neuropath). The neural path is the part of the physical path that directly produces consciousness.

And the awarepath is what the neural path produces: the stream of conscious experience itself.

Think of it as three concentric layers:

> **Physical path** = the body's complete physical journey through time. > **Neural path** = the brain's journey specifically. > **Awarepath** = the conscious experience that the brain produces.

The physical path includes the neural path, and the neural path determines the awarepath. But, and this is crucial, the relationship is not one-to-one.

The Overdetermination of Consciousness

Here is the chapter's central insight, and it is worth dwelling on.

Right now, as you read this sentence, a cosmic ray could strike a single atom in your liver and knock it out of position. Would your conscious experience change? Not by one iota. A bacterium in your gut could divide or not divide. Your blood sugar could be 95 mg/dL instead of 97. A skin cell on your left shin could be at a slightly different stage of its lifecycle. A capillary in your right foot could be carrying one more red blood cell or one fewer. None of these differences would change your awareness by even the faintest shade.

In fact, a thousand variations in your body's state — different blood sugar, different skin cell count, a bruise on your left knee versus no bruise, a particular atom in your liver being carbon-12 versus carbon-13, a different position of your left foot — would all produce the exact same moment of awareness. Your conscious experience would be identical across all of them.

This means that there are an enormous number of different physical paths, different complete physical trajectories of your body, that all produce the exact same neural path, and therefore the exact same awarepath. Your conscious experience is wildly **overdetermined** by the physical world. The space of body-states that produce your current moment of awareness is not a single point; it is a vast region, encompassing countless variations that make no difference to what you actually experience.

And it goes further. Even at the level of the brain, there is some degree of overdetermination. As we saw in the neuroscience chapter, neurons are noisy, synaptic transmission is probabilistic, and the brain's macro-level patterns are robust against minor variations in individual neural behavior. Not every microscopic difference in brain state produces a different conscious experience.

So your conscious identity, your awarepath, is not one thing. It is a class of things. It is the class of all physical arrangements that produce your conscious experience.

Branches and Possibilities

At every moment of your life, your future awarepath could branch in many different directions. If you turn left at the next intersection, you will have one sequence of experiences; turn right, and you will have a different sequence. Should a car run a red light at that intersection, your experiences will be different again.

But here is a subtlety that most people miss: many of these branches produce identical conscious experiences, at least for a while. Whether the air molecules to your left shift slightly one way or another makes no difference to what you experience. Whether a particular bacterium in your gut divides or not makes no difference to your consciousness. There are countless physical variations that have zero impact on your awarepath.

This means that from any given conscious moment, the space of possible future awarepaths fans out enormously — but many of those awarepaths are identical to each other. Your consciousness is overdetermined by physics: there are far more physical paths that produce your experience than there are distinct experiences.

And this overdetermination is what makes superimmortality possible. Because if your awarepath can be produced by a vast number of different physical paths, then even if one particular physical path ends — your body dies — there are, in principle, other physical paths that could produce the same continuation of your awarepath. The inner movie does not have to end just because one projector breaks down, if another projector somewhere is running the same film.

Your conscious experience is massively overdetermined by physics: countless variations in your body's physical state produce the same moment of awareness. If one physical path ends, others can in principle continue the same awarepath. But there is something deeper than the sequence of your experience — the quality that makes it yours, that persists even as the content changes from moment to moment. That quality needs a name.

CHAPTER 8

Ixperiencitness: What It's Like to Be You

Close your eyes for a moment and listen. Whatever you hear right now — traffic, wind, a refrigerator humming, silence — you are hearing it as you. Not as anyone else. Not as a generic conscious being. As the specific person you are, with your specific way of processing sound, shaped by every year of your life. The rumble of a passing truck lands differently in your awareness than it would in someone else's, because your brain parses it through a lifetime of accumulated associations, emotional tendencies, and cognitive habits that are yours alone.

That quality, that you-ness of your experience, is what this chapter is about.

The Sensepaducer

To see why this quality matters, and to understand what it really is, consider a thought experiment.

Imagine a hypothetical device called a **sensepaducer** — a machine that can generate any sequence of sensory signals and feed them directly into your brain. Suppose the neurons carrying information from your eyes, ears, skin, tongue, and nose to your brain were carefully disconnected and reconnected to this device instead. The sensepaducer now controls everything you see, hear, touch, taste, and smell.

Now imagine it feeds your brain sensory signals corresponding to having a completely different body. You look down and see unfamiliar hands — larger, smaller, darker, lighter, older, younger than your own. A different weight settles in your body, a different height, a different shape. In the mirror, a stranger's face stares back at you.

Here is the question: would you still be you?

Your conscious content would be dramatically different — you would be experiencing a different body, a different visual field, different tactile sensations. But think carefully. The way you react emotionally to what you see — the specific tinge of surprise, or curiosity, or unease — would still bear your signature. The way your attention moves, the way your thoughts form and connect, the particular flavor of your inner monologue: all of this would persist, because it arises not from the sensory signals coming into your brain but from the deep structure and functioning of the brain itself.

In other words, there is something about your experience that goes deeper than its content. Deeper than what you happen to see or hear or feel at any given moment. It is the way you experience, the cognitive and emotional architecture through which all experience passes. And that deeper quality is what we need a name for.

Naming the Concept

The word is **ixperiencitness**. It is built from the phrase "I experience it," with the suffix "-ness" added to turn it into a noun, just as "consciousness" is built from "conscious."

No existing term captures exactly what I mean, and precision matters more than elegance when the stakes are this high. The concept behind the word is one you already understand intimately, even if you have never had a name for it.

Ixperiencitness means, simply and precisely: **the quality of subjective experience that makes it *your* experience.**

It is not just what you are conscious of (the sights, sounds, thoughts, and feelings). Rather, it is the deeper quality of what it is like to experience them, the "you-ness" of your experience. The redness of red as you see it. The pain of pain as you feel it. The joy of joy as you live it. The ineffable sense that there is someone home, experiencing all of this — and that the someone is you.

The Concept You Already Know

You know what ixperiencitness is because you live inside it every moment of your life. You know what it is like to be you — that your experience has a particular quality, a "feel," that is yours. And every morning when you wake up, you recognize immediately that you are you and not someone else. This knowing, this intimate familiarity with the quality of your own experience, is your awareness of your own ixperiencitness.

The sensepaducer thought experiment revealed something important about it: ixperiencitness is not about the content of experience (what you see, hear, feel) but about the way you experience it, the deep cognitive and emotional patterns that make your experience distinctively yours. It is these deep patterns that are produced by the underlying architecture of your brain, and it is these patterns that survive changes in sensory content, in body, and even (as we shall argue) in physical substrate.

Now here is the key question: **What determines your ixperiencitness?**

The answer, consistent with everything we have established, is: the way your brain is organized and operating. Your ixperiencitness is produced by that physical arrangement and activity. It is not produced by your specific atoms (which are constantly being replaced), nor by a soul (for which there is no evidence), nor by the history of your body (which affects your consciousness only insofar as it shapes your present brain structure).

And if ixperiencitness is produced by the pattern, then **identical patterns produce identical ixperiencitness.**

This is the critical claim. If someone built a brain that was structurally and functionally identical to yours (same neural connections, same firing patterns, same chemical balances) the ixperiencitness it produced would be identical to yours. The person with that brain would actually experience the same thing (and would say so, too). Not something similar. Not something analogous. The same experience.

The Ixperiencitness Continuum

Here is a subtlety that is easy to miss but critically important: ixperiencitness is not an all-or-nothing concept. It is a continuum.

What does this mean in practice? Think of identical twins raised together. They share genetics, environment, and many formative experiences. When one twin hears their mother's voice, the warmth they feel, that specific, personal warmth, colored by a lifetime of shared memories, is probably very similar to what the other twin feels hearing the same voice. Not identical, perhaps, but deeply overlapping. Each twin processes that voice through a brain shaped by nearly the same forces. The way they experience it — the emotional texture, the associative echoes, the felt quality — overlaps substantially.

That overlap is what I mean by shared ixperiencitness.

Now consider two strangers from different cultures, different languages, different life histories. When they each hear a piece of music, their experiences may share some basic structural features — both perceive melody, rhythm, harmony — but the way they experience it, the personal associations, the emotional resonances, the cognitive habits through which the sound is filtered, will be far more different. Their ixperiencitness overlaps less.

Two consciousnesses do not have to be completely identical to share some degree of ixperiencitness. They can be partially similar. Your consciousness right now is not identical to your consciousness five years ago; you have different memories, different knowledge, perhaps different emotional tendencies. But your ixperiencitness is still recognizably yours. There is a continuity of ixperiencitness that persists despite changes in the specific content of consciousness.

Ixperiencitness has many different components: personality, emotional disposition, cognitive style, sensory processing, memory structure, and much more. Each of these components can vary independently, and each can match between two consciousnesses to a greater or lesser degree. Two consciousnesses produced by identical brains would have fully identical ixperiencitness. Consciousnesses produced by very similar brains — like those of identical twins — would have highly overlapping ixperiencitness. And those produced by moderately similar brains would overlap less. And so on, in a smooth gradient.

This continuum concept is important because it means that immortality is not an all-or-nothing proposition. A future consciousness with fully identical ixperiencitness to yours is unambiguously a case of your survival. But what about one with highly overlapping ixperiencitness? Or moderately overlapping? At what point does a future consciousness stop being "you"?

There is no sharp boundary, just as there is no sharp boundary between "you" at age 20 and "you" at age 80, even though your ixperiencitness has changed considerably over those decades. Identity is a spectrum, not a binary.

This is one of the most radical and most important insights of Ixperiencit Theory. It challenges our binary intuition that you either survive or you don't, and replaces it with a much more nuanced, and much more accurate, picture.

Why Existing Terms Fall Short

You might wonder why we need a new word when philosophers already have terms like "qualia" and "phenomenal consciousness." The answer is that ixperiencitness is more specific and more useful than either of these.

Qualia typically refer to individual sensory qualities — the redness of red, the painfulness of pain. But ixperiencitness refers to the total quality of a conscious experience: the entire experiential field of a conscious moment, far beyond any single sensation. It is the whole of what it is like to be you at a given moment, encompassing your sensations, emotions, thoughts, memories, sense of self, and everything else that constitutes your experience.

Phenomenal consciousness is closer, but it is typically used in philosophy to refer to the general phenomenon of subjective experience, the fact that "there is something it is like" to be a conscious being. Ixperiencitness is more specific: it refers to your particular subjective experience. It is what distinguishes your experience from someone else's.

To clarify how these concepts relate:

| Concept | Scope | Example | |---|---|---| | **Qualia** | Individual sensations | The redness of red | | **Phenomenal consciousness** | Subjective experience in general | The fact that "there is something it is like" to be a bat | | **Ixperiencitness** | Your personal experiential signature | What makes your experience of red yours, distinct from someone else's experience of red |

Most importantly, ixperiencitness is the concept we need for talking about survival, life after death, and immortality. Because the question of whether you survive death is ultimately a question about ixperiencitness: does some future conscious being experience the world with your ixperiencitness?

The question is no longer the vague "do you live on?" It is specific and, in principle, answerable: does some future conscious being experience the world with your ixperiencitness? Before we can follow that question to its conclusion, we need to face the deepest challenge to the framework — the hard problem of consciousness — and be honest about what it does and does not threaten.

CHAPTER 9

The Hard Problem — Honestly

I have been accused, more than once, of waving away the hard problem of consciousness. Of treating it as a minor technical detail rather than the profound mystery it is. Of building an entire theory of immortality on a foundation that, according to some of the smartest people alive, might have a crack running through it.

So let me face the hard problem directly, honestly, and without flinching.

What the Hard Problem Actually Is

In 1995, the philosopher David Chalmers drew a distinction that changed how we talk about consciousness.11 He separated the "easy problems" of consciousness from the "hard problem."

The easy problems (easy in principle, extremely difficult in practice) are about explaining the functions of consciousness: How does the brain integrate information from different senses? How does it direct attention? In what way does it control behavior? How does it distinguish wakefulness from sleep? These are all questions about what the brain does, and they are, in principle, answerable by the standard methods of neuroscience: identify the neural mechanisms, trace the circuits, build computational models.

The hard problem is different. It asks: **Why is any of this accompanied by subjective experience?** Why does it feel like something to process information? Why doesn't the brain just do all its information processing "in the dark," without any inner experience at all?

This is a genuinely puzzling question. Everything neuroscience tells us about the brain is, in principle, describable in terms of structure and function — neurons firing, chemicals diffusing, signals propagating. None of these descriptions, no matter how detailed, seem to entail that there is something it feels like to be the system doing the processing. You can describe every synapse, every ion channel, every neurotransmitter molecule, and it seems like you've left something out. The redness of red. The painfulness of pain. The what it's like of experience.

This is not a gap in our knowledge that will obviously be filled by more data. It is a gap in our framework. Our framework for understanding the physical world is built on structure and function. But subjective experience does not seem to be a structure or a function. It seems to be something else entirely, something that accompanies structure and function but is not identical to it.

Why This Matters for Ixperiencit Theory

The hard problem matters for this book because my entire argument rests on the claim that structure and functioning produce consciousness. If there is a genuine gap between structure/functioning and consciousness — if consciousness involves something beyond what physical description can capture, then my premises might be incomplete.

Specifically, two dangerous possibilities arise:

Danger 1: Extra ingredients. Maybe consciousness requires something in addition to the physical pattern: some psychophysical law, some fundamental property of matter, some organizing principle that current physics doesn't describe. If so, identical physical organization might not be sufficient for identical consciousness. The extra ingredient might differ between two identical brains, producing different experiences.

Danger 2: Multiple realizability fails. Maybe consciousness is more tightly bound to specific physical substrates than I have claimed. Maybe it's not enough for the pattern to be the same; maybe the specific material matters in ways we don't yet understand.

I want to take both of these seriously.

My Position: The Hard Problem Is Real but Irrelevant to the Argument

Here is my position, stated as carefully as I can.

The hard problem is a real philosophical puzzle. I do not claim to have solved it. I do not think anyone has solved it. It may be the deepest question in all of philosophy, and it may take centuries or millennia to answer — if it can be answered at all.

But, and this is the critical point, **the hard problem does not need to be solved for the three premises of Ixperiencit Theory to be true.**

Consider an analogy. Imagine you are an engineer in the 18th century. You do not understand how gravity works at a fundamental level. (Newton's theory describes what gravity does but not why mass attracts mass.) Despite this lack of fundamental understanding, you can still build bridges, predict eclipses, and launch cannonballs with remarkable accuracy. Your engineering works because you have a reliable, well-confirmed relationship between mass, distance, and gravitational force, even though you don't understand why this relationship holds.

The relationship between brain structure/functioning and consciousness is like this. We do not understand why certain physical processes are accompanied by subjective experience. This is the hard problem, and it is unsolved. But we have an extremely reliable, well-confirmed relationship between the two. Alter the brain, and consciousness changes in predictable ways. Restore the brain, and consciousness returns. The same brain structure produces the same behavioral reports of experience, consistently and reproducibly.

This empirical relationship is what my premises rest on, not on a solution to the hard problem. Premise One says that the brain produces consciousness. This is an empirical claim supported by overwhelming evidence. It does not require understanding how the brain produces consciousness, any more than using gravity requires understanding how mass produces gravitational attraction.

Premise Three says that identical physical organization produces identical consciousness. This is a prediction based on the most basic principle of physics: identical physical systems produce identical results. There is no known exception to this principle, and the hard problem does not provide one. The hard problem says we don't understand why brains produce consciousness. It does not say that identical brains produce different consciousness.

The Landscape of Possibilities

Let me map out the logical territory. There are four main positions on the hard problem, and I want to show that Ixperiencit Theory survives under three of the four:

Position 1: Physicalism (consciousness is structure and functioning).

If consciousness just is certain physical patterns, if the hard problem dissolves because consciousness turns out to be fully explicable in physical terms, then Ixperiencit Theory follows straightforwardly.

Identical physical organization produces identical consciousness because they are identical consciousness. This is the simplest case.

Position 2: Property dualism (consciousness accompanies the physical pattern).

If consciousness is a non-physical property that reliably accompanies certain physical structures, like a shadow accompanies an object, then Ixperiencit Theory still works, as long as the accompaniment is lawful. Identical physical structures cast identical "shadows." Identical brains produce identical consciousness. We don't know why consciousness accompanies brains, but we know that it does, and that the mapping is consistent. This is Chalmers's own view, and notice: it supports, rather than undermines, the claim that identical physical organization produces identical consciousness.

Position 3: Panpsychism (consciousness is fundamental to matter).

If consciousness is a fundamental property of matter, if even electrons have some proto-experiential quality, then the hard problem is dissolved differently: consciousness doesn't emerge from matter because it was there all along. In this case, the physical organization of a brain would determine what kind of consciousness is produced by organizing the fundamental experiential properties of its constituent matter. Identical organization would produce identical consciousness. Ixperiencit Theory survives.

Position 4: Strong dualism (consciousness is independent of the physical pattern).

If consciousness is produced by a non-physical substance (a soul, a mind-stuff, something entirely separate from the physical) then it is possible that identical physical structures produce different conscious experiences, because the soul is different. In this case, Ixperiencit Theory fails.

But Position 4 is the one with the least empirical support. There is no evidence for souls or mind-stuff. Every alteration of the brain that we have ever studied has produced a corresponding alteration in consciousness. There is no case in which brain structure was unchanged but consciousness changed (setting aside changes in functioning, which is part of our framework). Position 4 requires a radical departure from everything we know about the brain-consciousness relationship, and it requires positing entities (souls, mind-stuff) for which there is zero evidence.

An Honest Accounting

Let me be completely honest about what we know and what we don't.

What we know:

- Brain organization and activity are tightly correlated with conscious experience.
- Altering brain structure/functioning alters consciousness in predictable, consistent ways.
- Eliminating brain functioning (anesthesia, death) eliminates consciousness.
- Restoring brain functioning restores consciousness.
- Similar brains produce similar reports of conscious experience.

What we don't know:

- Why any physical process is accompanied by subjective experience.
- Whether consciousness can exist without any physical substrate.
- Whether the relationship between brain states and conscious states is identity (they are the same thing), supervenience

(consciousness is determined by but not identical to brain states), or something else entirely.
- What the minimal physical requirements for consciousness are.
- Whether consciousness is continuous or discrete.

What Ixperiencit Theory requires:

- That consciousness is produced by (or at minimum, reliably accompanies) the brain's physical organization and activity.
- That the relationship is consistent: identical organization always produces identical consciousness.

Notice what the theory does not require: a solution to the hard problem. It does not require understanding why brains produce consciousness. It only requires that they do, and that they do so consistently.

This is a modest epistemological requirement. It asks only that we trust the empirical evidence, the same evidence that lets us build MRI machines, develop anesthetics, and treat neurological disorders. If this evidence is trustworthy enough to base medical practice on, it is trustworthy enough to base a theory of consciousness on.

The Deepest Question

But I want to end this chapter with something I rarely say: I don't have all the answers.

The hard problem of consciousness is humbling. It reminds us that we are at the very beginning of understanding the most fundamental aspect of our existence. We are like astronomers in the age of Ptolemy: we can track the patterns with remarkable accuracy, but we do not yet understand the machinery that produces them.

What I am confident of is this: whatever the solution to the hard problem turns out to be, it will not overthrow the empirical evidence that consciousness is produced by the brain's physical organization. It may deepen our understanding of this relationship. It may reveal that consciousness is more fundamental than we thought, or that it involves principles we have not yet imagined. But it will not erase the evidence that damage to the brain damages consciousness, that identical brains produce identical behavior, and that the pattern is what matters.

The hard problem tells us that our understanding is incomplete. It does not tell us that our evidence is wrong. And the evidence points clearly to the three premises on which this entire book rests.

One final note. If a non-physical ingredient to consciousness were demonstrated, if Position 4 were somehow confirmed, I would revise the theory. Ixperiencit Theory is built on premises, not on dogma. If the premises are wrong, the conclusions are wrong. That willingness to revise is what separates this framework from the religious afterlife concepts it sometimes superficially resembles. We will specify exactly what would falsify the theory in Chapter 19.

CHAPTER 10

The Multiplicity of You

Now we arrive at one of the most radical and most important ideas in Ixperiencit Theory: **conscious multiplicity**.

If multiplicity sounds wrong to you, you are in good company. It violates one of our deepest intuitions, the feeling that there is exactly one "you," housed in exactly one body. But examine what the idea actually claims, and you may find it harder to reject than you expect.

The standard view of identity says: there is one you. You exist in one place, at one time, in one body. This is what we might call the **singularity** view of consciousness. One person, one consciousness, end of story.

Ixperiencit Theory proposes something far more interesting. There are potentially many instances of you, many physical systems that produce your conscious experience, spread across space and time. This is the **multiplicity** view.

Four Theories of Conscious Existence

To see where multiplicity fits among theories of identity, it helps to consider four possible theories:

1. Nonexistence.

There is no persistent self at all. Consciousness is an illusion, or at best a momentary flicker with no continuity. Each moment of experience is independent and unconnected to any other. This view has affinities with certain Buddhist philosophies that deny the existence of the self.

Ixperiencit Theory argues this goes too far. While it is true that the self is not what folk psychology imagines it to be, it is not nothing. Ixperiencitness is real (you really do experience being conscious) and it has continuity and structure over time. Denying the reality of the self throws the baby out with the bathwater.

2. Singularity.

There is exactly one you, tied to one body. When that body dies, you are gone forever. This is the common Western materialist view, and it is what most atheists and many scientists believe.

Ixperiencit Theory argues this is too narrow. Singularity follows only if you assume that consciousness is tied to a specific body rather than to structure and functioning. Once you accept that structure and functioning are what matter, singularity becomes an arbitrary and unsupported restriction.

3. Multiplicity.

There are many instances of you across space and time, many physical systems that produce your ixperiencitness. This is what Ixperiencit Theory proposes, and it is the view we will develop in this book.

4. Universality.

You are every conscious experience everywhere. All consciousness is one. This view has affinities with certain Hindu and panpsychist philosophies.

Ixperiencit Theory argues this goes too far in the other direction. While the physical pattern determines ixperiencitness, not every brain organization produces your ixperiencitness. A brain structured very differently from yours (a dolphin brain, or the brain of a person with a radically different personality and cognitive style) does not produce your ixperiencitness. Identity is real and has boundaries, even if those boundaries are fuzzy rather than sharp.

Multiplicity is the scientifically supported middle ground. You are not one body, but you are not everything. You are every physical system that produces your ixperiencitness.

Why Multiplicity Follows from Our Premises

To see why multiplicity follows from our premises, consider the sheer vastness of physical possibility.

The observable universe contains roughly 10^80 atoms. It has existed for about 13.8 billion years and will continue to exist for trillions of years (or possibly forever). In all that space and time, the number of possible arrangements of matter is staggeringly large, far larger than the number of atoms, because each atom can be in many different positions and states.

Among those unimaginably many arrangements, some will produce brain structures and functioning identical to yours. Given the scale of the universe and the precision required, this is a strong probability — though as we will see in Chapter 20, the exact likelihood depends on cosmological assumptions that remain uncertain.

These are not copies of you. They are you, in the only sense that matters: they produce your conscious experience. They have your memories, your personality. Each one looks out at the world through what feels like your eyes and is, from the inside, indistinguishable from you. Because the inside, the conscious experience, is determined by the physical pattern, and the pattern is the same.

The Challenge to Intuition

This challenges our intuition, because we are used to thinking of ourselves as tied to one particular body. But as we established in Chapters 4 and 5, the same pattern can exist in multiple places.

Think of it this way. Imagine that two identical brains exist simultaneously, in two different locations. Both are structured and functioning identically. According to our premises, both produce identical conscious experiences. Which one is "you"? The answer is: both. There is no principled way to say one is "real" and the other is a "copy," because there is no difference in the conscious experience. The experience is the identity, and the experience is the same.

This is exactly like two ocean waves with the same shape, amplitude, and frequency, moving through different stretches of water. Which wave is the "real" wave? The question is meaningless. Both are the same wave pattern, because a wave is not the water molecules it passes through; it is the pattern those molecules instantiate.

Multiplicity Through Time

Multiplicity encompasses more than multiple instances of you existing simultaneously. More importantly for the question of immortality, it includes instances of you existing at different times.

If, a thousand years from now, a brain comes into existence with organization identical to what yours has right now, that brain will produce your ixperiencitness. It will experience being you. From its perspective — which is your perspective, because the experience is the same — it will be a continuation of your consciousness.

And from your perspective, when it comes into existence, you will experience being it. Not because some kind of signal travels from your current brain to that future brain (there is no need for any such connection), but because the experience is identical, and experience is what you are.

This is multiplicity through time, and it is the foundation of superimmortality. Your consciousness does not end when your body dies. It continues whenever and wherever the right physical conditions produce the right pattern — which is to say, whenever and wherever the universe produces your ixperiencitness.

Now that the theoretical framework is in place, it is time to stress-test it — to push the premises to their extremes with thought experiments and see whether they hold.

PART THREE: THE THOUGHT EXPERIMENTS

CHAPTER 11

The Identical Replica

Let's make this concrete with a thought experiment.

Imagine that advanced technology creates an **identical replica** of you (what I call an identireplica in the full theoretical framework), an exact physical replica of your entire body, down to the atomic level. Every cell, every protein, every neural connection, every chemical concentration is identical. This replica is placed in an identical environment.

What happens?

The replica behaves exactly like you. It has your memories. It speaks like you, moves like you, thinks like you. If you ask it questions about your life, it answers correctly, not because it has been programmed with your biography, but because it has your brain, with your neural connections that encode your memories.

Does this replica experience consciousness? Of course. It has a fully functioning brain, structurally identical to yours. If your brain produces consciousness, so does this one. The structure and functioning are identical, so the consciousness is identical.

Now the critical question: **does the replica experience being you?**

Here is where most people hesitate. Their instinct says no — that's a copy, not me. But examine that instinct carefully. What exactly would be different about the replica's experience? It has your memories of your childhood and remembers reading this book. It feels like itself, like you, because its brain is producing the same ixperiencitness (recall: the quality of subjective experience that makes it yours, the "you-ness" of consciousness) as yours.

If you could somehow swap places with the replica (if you could experience the world through its brain instead of yours) you would notice nothing different. The experience would be identical, because the brain producing it is identical.

The objection that it's "just a copy" assumes that something about your specific atoms, beyond their arrangement and functioning, contributes to your consciousness. But as we established in Part One and Part Two, the evidence consistently points the other way.

The "It's Not Me" Response Revisited

When people say "a replica is not me," they are usually imagining the original and the replica existing side by side. Two bodies in the room, both claiming to be the same person. Surely only one can be "real."

But the correct analysis is subtler. Both you and the replica have identical ixperiencitness at the moment of creation. There are not two different conscious experiences happening; there are two instances of the same conscious experience, produced by two different physical systems — like lighting a second candle from the flame of the first.

From the moment of creation onward, the two instances will diverge as they receive different sensory inputs. But the ixperiencitness, the deep quality of "you-ness," remains the same even as the specific content of consciousness diverges.

Why This Matters for Immortality

The identical replica thought experiment matters because it establishes a key principle: **your conscious experience can be produced by a body other than your current one.**

If this principle is accepted, then the barrier to immortality falls. Death destroys one body, but it does not destroy the possibility of another body producing the same ixperiencitness. And wherever and whenever that ixperiencitness is produced, you are there. You are experiencing it. You are alive.

This is not resurrection or an afterlife. It is the logic of structure and functioning, applied consistently. (If you find yourself reaching for objections about causal continuity, about what makes the "original" special, about whether this is really testable — hold that thought. Part Four is devoted entirely to engaging the strongest counterarguments.)

CHAPTER 12

The Continuing Replica and the Paradox It Creates

The identical replica thought experiment establishes that your consciousness can be produced by another body. But a snapshot in time is not enough. What matters for immortality is continuity: not continuity of the body, but continuity of conscious experience.

This is where the **cidentireplica** comes in.

A cidentireplica (a **continuing identical replica**) is a body that not only starts with your physical organization but continues to match you over time. It doesn't just match you at one moment; it matches you moment by moment, producing the same awarepath you would produce.

How is this possible? In principle, it could be achieved in several ways: by placing the cidentireplica in an identical environment, by directly controlling its neural activity to match yours, or simply by letting it evolve naturally in conditions similar enough that its brain dynamics remain similar to yours.

The Cidentireplica Paradox

Here is the paradox that gives this thought experiment its power.

The cidentireplica, by definition, has behavior identical to yours. It says the same things, does the same things, responds to questions in the same way. If you ask it its name, it gives your name. If you ask it about events from your past, it answers correctly (from its perspective, and its perspective is identical to yours).

No observer, no matter how sophisticated, no matter how much information they have, can tell the difference between you and the cidentireplica based on behavior alone. The behaviors are identical because the brains producing them are identical.

Now apply this to an analogy. Imagine two houses built from the same blueprint — same dimensions, same materials, same layout. You walk through both. The experience of living in one is indistinguishable from the experience of living in the other. The rooms are the same size, the doors open the same way, the light falls through the windows at the same angles. They are the same house — not "similar houses," not "faithful copies," but the same design, instantiated twice.

Now imagine someone claiming: "Even though both houses are identical in every measurable way, only this house is the real house. The other one might look and feel the same, but it's not. Not really."

This claim is clearly absurd when applied to blueprints and houses. Why should it be any less absurd when applied to brains and consciousness?

The cidentireplica paradox forces us to choose: either identical structure and functioning produces identical consciousness (which implies multiplicity and immortality), or there is some invisible, undetectable "extra ingredient" that distinguishes one consciousness from another even when all physical properties are the same (which is indistinguishable from belief in a soul).

This is a stark choice, and the burden of proof falls heavily on the side claiming an invisible extra ingredient.

What the Cidentireplica Tells Us About Death

Now take the next step. The cidentireplica has a continuing existence. Its brain continues to function identically to yours, moment by moment. It continues to produce the same conscious experience you would produce.

If your original body dies, and the cidentireplica continues — **from the perspective of conscious experience, nothing has ended.** The same ixperiencitness continues to be produced. The pattern continues. You continue, in every way that consciousness can measure.

From your perspective (if you were the one who died), you would not experience death. You would not experience a gap. You would simply continue having experiences, produced by the cidentireplica's brain, without ever knowing that your original body had ceased to exist. This is because consciousness is produced by the physical pattern, and the relevant pattern never stopped.

From the perspective of outside observers who knew about both bodies, they would see one body stop and another continue. They might say "the original died and only the copy survived." But from the experiential perspective — the only perspective that matters for the question of survival — there was no death. There was no interruption. There was just continuous conscious experience.

If a cidentireplica continues after your original body dies, then from the inside — from the only perspective that matters for the question of survival — conscious experience simply continues without interruption.

CHAPTER 13

The Teleporter Problem

Science fiction has given us one of the best thought experiments for understanding identity and consciousness: the teleporter.

In a typical science fiction teleporter (think Star Trek), your body is scanned at the molecular level, the information is transmitted to a distant location, and a new body is assembled from local matter using that information. You step into the teleporter on Earth and step out on Mars a moment later. Simple, right?

But here is the question that science fiction usually glosses over: **What happens to the original?**

Three Scenarios

Scenario 1: Destructive teleportation.
The original body is destroyed during the scanning process. A new body is assembled at the destination. This is the standard science fiction version.

Is the person who steps out on Mars the same person who stepped in on Earth? If consciousness is produced by structure and functioning, and the Mars body has the same structure and functioning as the Earth body did, then the answer is yes. The person on Mars has your memories, your personality, your ixperiencitness. From their perspective, they stepped into a booth on Earth and stepped out on Mars. They experienced no gap, no death, no discontinuity.

Scenario 2: Non-destructive teleportation.

The original body is not destroyed. It continues to exist on Earth while a copy is assembled on Mars.

Now there are two of you. The person on Earth has your memories and your ixperiencitness. The person on Mars has your memories and your ixperiencitness. Both claim to be you. Both are you, in terms of ixperiencitness.

This is conscious multiplicity in action. And it raises a disturbing question: if the person on Mars is you when the original survives, then the person on Mars is equally you when the original is destroyed. The Mars person's identity doesn't depend on what happens to the Earth person.

Scenario 3: The delayed-destruction version.

The original body is scheduled to be destroyed one hour after teleportation. During that hour, both the original and the copy exist.

During that hour, who is "you"? Both. You are experiencing two sets of experiences simultaneously, through two different bodies, in two different locations. From the outside, there are two people. From the inside, there are two instances of the same ixperiencitness, rapidly diverging as they have different experiences.

When the original is destroyed after one hour, one set of experiences ends and the other continues. From the perspective of the continuing body on Mars, nothing special happens at that moment; they don't feel the original's death. From the perspective of the original on Earth, they experience one hour of continued existence and then... nothing. Not death, exactly, because there is no experience of non-existence. Just a transition from one set of experiences to no experiences — experienced as nothing at all.

What Teleportation Teaches Us

The teleportation thought experiment teaches us several important lessons:

First, the identity of the teleported person does not depend on what happens to the original. If the Mars person is you (which they are, based on ixperiencitness), they are you regardless of whether the Earth person survives, is destroyed immediately, or is destroyed an hour later. Identity is determined by the consciousness being produced, not by the fate of other bodies.

Second, conscious multiplicity is a natural consequence of the physics of consciousness. Whenever two or more physical systems produce the same ixperiencitness, multiplicity exists. There is nothing paradoxical about this, any more than it is paradoxical for two rivers to carve the same winding pattern through similar terrain.

Third, teleportation without multiplicity is difficult to maintain on purely physical grounds. If you insist that the teleported copy is not really you — that only the original is really you — then you must explain what makes the original special. As we have seen, the most natural answer points toward some kind of non-physical essence. The strongest version of this objection is Parfit's causal-continuity requirement: that identity demands an unbroken causal chain from one moment of consciousness to the next. But notice that the teleporter does preserve a causal chain; the scanning process causally determines the reconstruction. The information that specifies the structure flows continuously from original to replica. What is broken is material continuity, not causal continuity. And as we have already seen, material continuity is broken by ordinary metabolism anyway.

Fourth — and this is the lesson most relevant to immortality — **if teleportation works, then so does resurrection.** Both involve creating a body with the right physical organization to produce your ixperiencitness. The only difference is that in teleportation, the creation happens during the original's lifetime, while in "resurrection," it happens after the original's death. But the mechanism is the same, and the result is the same: the production of your conscious experience by a new body.

Why the Original Need Not Worry

Here is a thought that might trouble you: if the teleporter creates a copy that is equally "you," then stepping into a destructive teleporter is effectively committing suicide while a copy of you walks away.

But this worry is based on the singularity view of identity, the view that there is exactly one "you," located in one body. If you accept multiplicity, the worry dissolves.

You are not your body. You are the ixperiencitness that your body produces. When you step into the teleporter, you step out at the other end. The body on Earth is destroyed, but you — the conscious experience — continue on Mars. There is no suicide, because there is no loss of ixperiencitness. The same ixperiencitness continues to be produced, just by different atoms in a different location.

If this still feels wrong, ask yourself: is there a difference between destructive teleportation and the normal process of atomic replacement that your body undergoes every day? In both cases, the atoms are replaced and the pattern continues. The only difference is the speed. And as we established in Chapter 4, the speed of matter replacement is irrelevant.

The Teleporter Litmus Test

Here is a useful exercise. Ask yourself honestly: **Would you step into a destructive teleporter?**

If yes, if you would accept that the person who steps out on Mars is you, then you have already accepted the core premises of Ixperiencit Theory. You have accepted that identity follows structure and functioning, not specific atoms or bodily continuity. And if identity follows structure and functioning, then superimmortality follows logically.

If no, if you insist that the person on Mars is "just a copy," then ask yourself what makes the original special. What property does the original have that the copy lacks? If you cannot identify a physical property (and you cannot, because they are physically identical), then your resistance is based on intuition, not evidence. And as we discussed in Chapter 3, intuitions about identity are among the least reliable intuitions we have.

The teleporter is a litmus test, and the neutral ground is narrower than it looks. Either way, it points to the same conclusion as the identical replica and the continuing replica: identity follows the physical pattern. But we do not need science fiction to see this principle at work. Something much more familiar demonstrates it every time a patient counts backward from ten.

CHAPTER 14

The Anesthesia Argument

One of the most powerful everyday demonstrations that consciousness is produced by brain functioning, and that it can be interrupted and restored, is the experience of general anesthesia.

When a person undergoes general anesthesia, their consciousness appears to be completely eliminated, suppressed below the threshold of detectability. There is no experience of passing time, no dreaming (though studies using the bispectral index (BIS) and isolated forearm technique have found evidence of residual neural processing under certain protocols), and no awareness of the surgical room. From the patient's perspective, there is absolutely nothing.

And then, when the anesthetic agents are metabolized and the brain resumes normal functioning, consciousness returns. The patient wakes up. They are themselves again: same memories, same personality, same ixperiencitness. The gap in consciousness might have been minutes or hours long, but from the patient's perspective, it was instantaneous. One moment they were counting backwards from ten; the next moment, they were waking up in the recovery room.

What Anesthesia Teaches Us

Anesthesia teaches us several things that are crucial for our argument:

First, consciousness is not a permanent property of the body. A body can exist — can even be alive, with a functioning heart and circulating blood — without producing consciousness. Consciousness is something the brain does when it functions in the right way, not something the brain has as an intrinsic property. Stop the right functioning (with anesthetic drugs), and consciousness stops. Restore the right functioning, and consciousness returns.

Second, gaps in consciousness do not affect identity. You are the same person after anesthesia as before, despite the gap. Your ixperiencitness is unchanged, your memories (up to the point of anesthesia) are intact, and your personality is the same. The gap, from the perspective of your conscious identity, might as well not have happened.

Third — and this is the key insight — **if a gap in consciousness does not affect identity, then it does not matter how long the gap is.** Whether the gap is one hour (surgery), eight hours (sleep), or a billion years (death followed by the eventual reproduction of your brain's structure and functioning), the result is the same from the experiential perspective: you go from one moment of consciousness to the next, with the gap simply not existing in your experience.

The Anesthesia Extension Argument

Now let me extend the anesthesia argument in a way that reveals its full power.

Imagine you are put under general anesthesia. While you are unconscious, a surgeon makes changes to your brain. Perhaps they add some neurons, alter some connections, modify the balance of neurotransmitters. When you wake up, you are slightly different — perhaps a bit smarter, or with slightly different emotional responses, or with a few memories added or removed.

Are you still you? Most people would say yes, especially if the changes are small. You are a modified version of yourself, but still recognizably you — still possessing your ixperiencitness, perhaps slightly altered.

Now imagine that the changes are larger. The surgeon rewires significant portions of your brain, to the point that when you wake up, you have some different memories, different personality traits, a different emotional character. You still have some continuity with your pre-surgery self, but you are substantially changed.

Are you still you? Here the answer gets murkier — you are less clearly the same person, because the ixperiencitness has changed more significantly. You are on a continuum between the original you and a genuinely different person.

Now push this to the extreme. Imagine the surgeon completely restructures your brain while you are under anesthesia, to the point that when you wake up, you have the brain organization of a completely different person, let's say, a person named George. You wake up with George's memories, George's personality, George's way of experiencing the world. In every way that matters, you believe you are George — and you have George's ixperiencitness.

What has happened? Your body — the same physical body that was wheeled into surgery — is now producing a completely different ixperiencitness. The original "you" is gone, not because the body was destroyed, but because the pattern that produced your ixperiencitness was replaced by a pattern that produces George's ixperiencitness.

This demonstrates, powerfully, that you are not your body. You are the pattern that produces your ixperiencitness. Change that pattern enough, and "you" disappear from your own body, even though the body continues to exist. Your body is just the vessel. The pattern is the self.

And if the self is the pattern, then the self can exist in any vessel that has the right organization.

CHAPTER 15

The Philosophical Zombie Meets the Ixperiencitness Zombie

The philosophical zombie (or "p-zombie") is one of the most famous thought experiments in the philosophy of mind. It imagines a being that is physically identical to you in every way (same atoms, same structure, same functioning) but that has no conscious experience at all. It behaves exactly like you, responds to stimuli exactly like you, but there is "nobody home." The lights are on, but nobody is in.

Philosophers use the p-zombie to argue about the nature of consciousness. Can we conceive of a p-zombie? If so, does that prove that consciousness is something over and above the physical pattern?

Ixperiencit Theory has a clear answer to this question, and it also introduces a new and more interesting variant: the **ixperiencitness zombie**.

The Standard Zombie

Can a body with identical structure and functioning to yours be completely unconscious? Ixperiencit Theory says no. If our premises are correct (if structure and functioning produce consciousness) then identical physical organization must produce identical consciousness. A p-zombie is logically impossible given our premises, not a matter of practical difficulty but of logical contradiction.

Chalmers disagrees, and his argument deserves respect. His position is that we can coherently conceive of a p-zombie, that there is no logical contradiction in imagining a physically identical being that lacks experience. If this conceivability is genuine (and not just a failure of imagination), then it follows that consciousness is not logically necessitated by physical structure. There must be something more, some psychophysical law or fundamental property, that links the physical to the experiential.

Here is where Ixperiencit Theory parts company with Chalmers, but perhaps less dramatically than you might expect.

Ixperiencit Theory agrees that we lack a complete explanation of why physical structure produces consciousness. It agrees that the hard problem is real and unsolved. What it disputes is the inference from conceivability to possibility. We can "conceive" of water that is not H2O; we can imagine a clear, tasteless liquid that quenches thirst but has a different molecular formula. But water is H2O, necessarily, and our ability to imagine otherwise reflects a limit on our imagination, not a fact about water. Similarly, our ability to imagine a p-zombie may reflect a limit on our understanding of the consciousness-structure relationship, not a genuine possibility.

More importantly for our purposes: even if Chalmers is right that consciousness involves something beyond physical structure, his own view (property dualism) holds that consciousness lawfully accompanies physical structure. Identical structure, identical consciousness, not because consciousness reduces to structure, but because the psychophysical laws are consistent. If so, all three premises of Ixperiencit Theory survive, and the argument for superimmortality goes through.

The p-zombie debate is a deep one, and I do not claim to have settled it here. (We engage with Chalmers's full position more carefully in Part Four, where we steelman the strongest objections to the theory.) What matters for our purposes is this: whether consciousness reduces to structure and functioning (physicalism) or accompanies structure and functioning (Chalmers's property dualism), identical structure and functioning produces identical consciousness. And that is all the argument requires.

The Ixperiencitness Zombie

But Ixperiencit Theory introduces a more subtle and more interesting concept: the **ixperiencitness zombie**.

An ixperiencitness zombie is a being that is conscious — it has subjective experience — but that does not have your ixperiencitness. It experiences being conscious, but it does not experience being you. It has somebody home, but that somebody is not you.

Is this possible? And if so, when does it happen?

Consider a body with physical organization similar to yours but not identical. It might be a brain with slightly different neural wiring, or different neurotransmitter levels, or a different pattern of synaptic strengths. This brain produces consciousness — but it produces a consciousness with a different ixperiencitness than yours.

From the outside, this being might be indistinguishable from you (if the differences in brain structure are small enough that they don't affect behavior). But from the inside, the experience is different, not yours.

The ixperiencitness zombie concept highlights the importance of the distinction between behavior and experience. Two beings can behave identically while having different ixperiencitnesses. This is because the mapping from brain structure to behavior is many-to-one: many slightly different brain structures can produce the same behavior. But the mapping from brain structure to ixperiencitness may be more sensitive; slight differences in structure might produce different ixperiencitnesses even when the behavior is the same.

This has important implications for the science of superimmortality. It means that accurately reproducing your ixperiencitness requires more than just reproducing your behavior. It requires reproducing the brain's underlying architecture with sufficient fidelity that the same ixperiencitness is produced. How much fidelity is "sufficient" is a scientific question that we do not yet have the tools to answer. But the principle is clear.

CHAPTER 16

The Test for Originalness

Here is one more thought experiment that reveals the deep implications of our theory.

Imagine that "all-knowing" experimenters (scientists with perfect knowledge of physics and unlimited technological capability) create several cidentireplicas of you. They now have, let's say, five bodies in a room: the original you and four cidentireplicas. All five have identical structure and functioning, behave identically, and claim to be the original. Each one has memories of being the original. Every one of them believes, sincerely and with full conviction, that they are the "real" one.

The experimenters know the complete physical history of all five bodies — they know which one has been continuously existing since birth and which four were recently created. But here is the challenge: **can they determine which is the original without consulting that historical record?**

The answer is no. Based on present physical properties alone — structure, functioning, behavior, conscious reports — the five bodies are indistinguishable. There is no physical test, no measurement, no observation that can identify the original based solely on its current state.

This means that "originalness" is not a physical property. It is a historical label — a fact about the past, not about the present. And if consciousness is determined by the present state of the brain (its current organization and activity), then originalness is irrelevant to consciousness.

What This Tells Us

The test for originalness reveals something important: **being the "original" confers no special status in terms of conscious experience.**

The original and the cidentireplicas all have the same ixperiencitness. They all have the same memories. They all have the same sense of being "themselves." The only difference between them is their history — which specific atoms were used, how long the body has been in existence, where it has been. And none of these historical facts affect the present conscious experience.

This completely undermines the common objection to multiplicity: "But the original is the real one, and the copies are just copies." If you can't tell the difference — if even omniscient experimenters can't tell the difference based on present physical properties — then the distinction is not physically real. It is a prejudice, born of folk psychology, with no basis in science.

An analogy: imagine you have five identical computers, all running the same program, all producing the same output. One of them was manufactured in January, and the other four were manufactured in February. Can you tell which was made first by looking at the computers and their outputs? Of course not. The date of manufacture is a historical fact that has no effect on the computer's present state or operation. The January computer is not the "real" computer. It is just the oldest.

The same applies to bodies and consciousness. The "original" body is not the "real" you. It is just the oldest instance. All instances that produce your ixperiencitness are equally "you."

And this means that if the original body dies while the cidentireplicas continue, you have not died. You have lost one instance, but the others continue. Your ixperiencitness is still being produced. You are still alive — in every sense that matters.

The Deeper Lesson

The test for originalness reveals something that goes beyond the debate about replicas. It reveals that the concept of "originalness," which feels so fundamental to our sense of identity, has no physical basis. It is a label we attach to things, not a property things have.

This matters because the fear of death is, at its core, a fear about the loss of originalness. "When I die, the original me is gone." But if originalness is not a physical property, if it cannot be detected, measured, or even defined in terms of the present state of a body, then it cannot be the thing that makes your consciousness yours. And if it is not what makes your consciousness yours, then losing it (in death) does not mean losing your consciousness.

What makes your consciousness yours is ixperiencitness, and ixperiencitness, unlike originalness, is a property of the present physical state. It can, in principle, be measured (once we develop the right instruments). And it can be reproduced.

CHAPTER 17

Experience Machines and Designed Consciousness

Imagine a device that could feed any sequence of sensory signals directly to your brain, bypassing your eyes, ears, skin, and all other sense organs. This device, which Ixperiencit Theory calls a **sensepaducer**, could make you experience anything: a tropical beach, the surface of Mars, a world that has never existed, a body that is not your own.

From the inside, the experience would be indistinguishable from reality. Your brain would process the artificial sensory signals exactly as it processes real ones, because it has no way to tell the difference. A signal from a sensepaducer and a signal from a real eye hit the same neurons, trigger the same cascades, produce the same conscious experience.

The sensepaducer is a thought experiment, but it is also a technology that is approaching reality. Virtual reality headsets already deliver artificial visual and auditory signals to the brain. Brain-computer interfaces are beginning to deliver signals directly to neural tissue. The full sensepaducer — controlling all sensory input to the brain — is a matter of engineering, not of principle.

And the implications for understanding consciousness, identity, and immortality are significant.

The Experience Machine Thought Experiment

The philosopher Robert Nozick proposed a famous thought experiment in 1974: the **Experience Machine.**12 Imagine a machine that could give you any experience you desired. You could experience being a great artist, a beloved leader, a tireless explorer, whatever you choose. While connected to the machine, you would have no idea you were connected; the experiences would feel completely real.

Would you plug in?

Nozick expected most people to say no, and he used this intuition to argue that we value more than just subjective experience; we value actually doing things, actually being things, actually living in the real world.

Ixperiencit Theory has a more careful take on this thought experiment.

What the Sensepaducer Reveals

The sensepaducer thought experiment is not primarily about whether you would choose to plug in. It is about what the possibility of a sensepaducer tells us about the nature of consciousness.

Lesson 1: Consciousness depends on brain processing, not on external reality.

If a sensepaducer can produce indistinguishable experiences, then your conscious experience is not directly caused by the external world. It is caused by your brain's processing of signals — and those signals can come from the real world or from a machine. What matters for consciousness is the brain's internal state, not its external cause.

This reinforces what we established earlier: consciousness is an internal state of the brain, not a direct report on external reality.

Lesson 2: "Real" experience and "artificial" experience are the same experience.

If you cannot tell the difference from the inside — if the sensepaducer experience and the real experience produce identical brain states — then they are, in terms of consciousness, the same experience. The distinction between "real" and "simulated" is meaningful from the outside but meaningless from the inside.

This has staggering implications. It means that any conscious experience you have ever had — any joy, any love, any moment of beauty — could have been produced by a brain in a vat receiving sensepaducer signals. And it would have been just as real, just as meaningful, just as truly yours, because the consciousness was the same.

Lesson 3: The space of possible experiences is far larger than the space of actual experiences.

Your current sensory organs limit you to a narrow slice of possible experience. You see visible light, hear a narrow range of sounds, feel only certain physical stimuli. A sensepaducer is not bound by these limitations. It could deliver signals that correspond to seeing infrared, hearing ultrasound, feeling magnetic fields, senses that human bodies don't naturally have.

But it could go further. It could deliver signals that correspond to no known sense at all, signals that activate neural patterns no natural stimulus has ever triggered, producing experiences that no human has ever had. Entirely new colors. Entirely new emotions. Entirely new ways of being conscious.

This expands the scope of superimmortality enormously. Your future existences need not be limited to the kinds of experience that current human biology can produce. With sensepaducer technology, the space of possible conscious experiences becomes, for practical purposes, infinite.

Awarepaducers: Designing Consciousness from Scratch

The sensepaducer concept is grounded in technology that already exists in primitive form. The awarepaducer, which we turn to now, is a much larger leap; it describes a technology that does not yet exist and may be generations away. But the concept follows logically from the same premises, and thinking through its implications now helps clarify what the theory actually predicts.

Ixperiencit Theory goes beyond the sensepaducer to propose an even more powerful concept: the **awarepaducer**. While a sensepaducer controls the inputs to a brain (the sensory signals), an awarepaducer controls the entire state of a brain-like system — its structure, its functioning, and therefore its consciousness.

An awarepaducer doesn't just give you a new experience. It gives you a new self. By controlling the structure and functioning of a brain-like system, an awarepaducer can produce any ixperiencitness — any conscious experience, belonging to any possible being.

With awarepaducer technology, the following becomes possible:

- **Reproducing any past consciousness.** If we have sufficient information about the organization of a past brain (through brain scanning, historical records, or inference), an awarepaducer could reproduce that consciousness, bringing a past person back to experiential life.
- **Creating enhanced consciousnesses.** An awarepaducer could produce consciousnesses that are smarter, more perceptive,

more emotionally rich, or more creative than any natural brain has ever produced.

- **Exploring the space of consciousness.** By systematically varying the parameters of a brain-like system, an awarepaducer could explore the vast space of possible consciousnesses, discovering new forms of experience that evolution never produced.
- **Merging consciousnesses.** An awarepaducer could produce a consciousness that combines aspects of two or more different ixperiencitnesses, blending elements of multiple identities into a single experience. This could allow a form of shared identity that is currently impossible.

The Ethics of Designed Experience

The technology of designed consciousness raises ethical questions that humanity has never faced before.

Who decides what consciousnesses to create?

If awarepaducer technology becomes available, someone must decide how to use it. The potential for creating positive, enriching, fulfilling conscious experiences is matched by the potential for creating suffering, confusion, or horror. The ethics of consciousness design is one of the most important problems that future generations will face.

Do designed consciousnesses have rights?

If a sensepaducer creates a conscious being that believes it is living a real life, does that being have moral status? Ixperiencit Theory says yes: any being with ixperiencitness has moral status, regardless of how that ixperiencitness was produced. A consciousness designed by an awarepaducer is just as real — just as valuable, just as deserving of moral consideration — as a consciousness produced by biological evolution.

Is a designed paradise really paradise?

Nozick's intuition was that a life of designed experiences is somehow less valuable than a life of real ones. But Ixperiencit Theory challenges this intuition. If the consciousness is the same, if the experience of beauty, love, accomplishment, and meaning is identical whether it comes from "real" events or designed signals, then the distinction between "real" and "designed" may be less important than we intuitively feel.

This does not mean we should all plug into experience machines. There are good reasons to value engagement with the real world: the unpredictability, the challenge, the connection to other conscious beings. But these reasons are about the kind of experience that engagement with reality produces, not about some metaphysical distinction between "real" and "fake" experience.

What This Means for Your Future

If the premises hold and the technology develops along the trajectory outlined above, the implications for your future existences are transformative. We are now in the most speculative territory of this chapter, but it is speculation that follows directly from the argument, and I think it is worth following to its conclusion.

The technologies of designed consciousness, sensepaducers and awarepaducers, are among the most important developments that could shape the future of superimmortality.

Without these technologies, your future existences depend on the natural or coincidental production of brains similar to yours — a process that is uncertain and largely beyond anyone's control.

With these technologies, your future existences can be designed — crafted by intelligent beings who understand consciousness and who deliberately choose to produce specific ixperiencitnesses, including yours. The reproduction of your consciousness becomes not a cosmic accident but a deliberate act, guided by science and ethics.

This is why the science of consciousness matters so much. Every advance in understanding how physical organization produces experience brings us closer to the technology of designed consciousness. And every step toward that technology is a step toward a future in which superimmortality is a practical reality.

The sensepaducer and awarepaducer are more than thought experiments. They are blueprints for the future of consciousness, technologies that could transform superimmortality from a cosmic accident into a deliberate act. With them, the reproduction of your consciousness becomes a matter of science and engineering rather than chance. But before we look forward, we need to look back — at the strongest objections to everything we have argued so far, stated as powerfully as possible, and answered honestly.

PART FOUR: THE STRONGEST OBJECTIONS

CHAPTER 18

Steelmanning the Opposition

If you have read this far, you may be convinced. Or you may be deeply skeptical. Either way, you deserve to see the strongest arguments against the theory presented in this book: not straw men, not caricatures, but the real thing. The best philosophy of mind has produced several powerful objections to the kind of view I am defending, and I owe it to you (and to intellectual honesty) to engage with them directly.

Let me steelman the opposition.

Objection One: The Causal-Continuity Requirement (Parfit's Challenge)

The philosopher Derek Parfit, in his landmark book Reasons and Persons13, spent hundreds of pages analyzing exactly the kind of thought experiments we explored in Part Three. Parfit was sympathetic to the idea that identity is about psychological continuity (memories, personality, beliefs) rather than about some unchanging soul. In many ways, his work paved the road this book travels on.

But Parfit insisted on something this book does not: **causal continuity**. For Parfit, it matters not just that your memories and personality exist in some future brain, but that they got there through the right kind of causal process. A replica that has your memories because it was built from a blueprint of your brain is, in Parfit's view, fundamentally different from a being that has your memories because it inherited them through normal biological and psychological processes, through the unbroken chain of cause and effect that started with your birth.

Why might causal continuity matter? Here is the strongest version of the argument.

Consider two scenarios. In the first, you go to sleep and wake up the next morning. Your memories, personality, and ixperiencitness persist because your brain maintained its structure overnight: neurons stayed connected, synaptic weights were preserved, the causal machinery of your biology kept everything in place. There is an unbroken physical chain linking last night's brain to this morning's brain.

In the second scenario, you are destroyed in your sleep, and a replica is built from scratch in the same bed. This replica has all your memories, all your personality traits, and, according to our premises, the same ixperiencitness. But there is no causal chain linking you to the replica. The replica's memories did not come from your experiences. They were manufactured.

Parfit (and many philosophers after him) would say these are genuinely different situations. In the first case, you survived. In the second, you died, and someone new was created who merely thinks they are you. The difference is not in the content of the conscious experience but in the causal history that produced it.

My response: This is the most serious philosophical objection to Ixperiencit Theory, and I want to address it carefully rather than dismiss it.

First, note what the causal-continuity requirement actually claims: that two brains can be physically identical in every measurable way (same structure, same functioning, same ixperiencitness) and yet one of them is "really you" while the other is not. The difference lies entirely in the history of how each brain came to be in its current state.

But here is the problem: history is not a property of the present physical state. If two brains are identical right now (atom for atom, connection for connection, firing pattern for firing pattern) then no measurement you could perform on them would reveal which one has the "right" causal history. The causal history is a fact about the past, not about the present. And consciousness is produced by the present state of the brain.

To see why this matters, consider a more extreme version of the thought experiment. Suppose you learn that last night, while you slept, advanced nanotechnology replaced every atom in your brain with an identical atom (same element, same position, same quantum state). The causal chain was broken at the atomic level, even though the structure was preserved. Are you the same person this morning? I think almost everyone would say yes. The "causal chain" between last night's specific atoms and this morning's specific atoms was broken, but the structure and functioning were maintained, and that is what matters.

Now push further. Suppose the nanotechnology didn't just replace atoms; it disassembled your brain entirely and reassembled it from scratch, using the same blueprint. The structure is identical. The functioning is identical. But there was a moment when your brain did not exist. Is this different from normal sleep, where your brain maintained its structure continuously?

If consciousness is produced by the physical pattern, and the pattern is the same, then the consciousness is the same. The causal-continuity objection asks us to believe that an invisible, undetectable property, the "right kind of history," makes a difference to consciousness that no physical measurement can reveal. This is structurally identical to believing in a soul. It posits something beyond the physical that determines identity.

There is a more sophisticated version of this objection that deserves a direct response. A philosopher might say: "I agree that history is not a present physical property. But identity is not a present physical property either; it is a relation that holds between temporal stages of a person. And relations can depend on causal history even when present states don't."

The analogy often used is the art forgery. A forged painting and a genuine Vermeer can be molecule-for-molecule identical. The forgery has every physical property the genuine article has. But "being a genuine Vermeer" is not a physical property of the canvas; it is a relational, historical property. The forgery is not a Vermeer despite being physically identical, because the causal history is wrong.

This is a clever argument, and it deserves a careful answer rather than a dismissal. Here is mine: consciousness is not like authenticity. Whether a painting is "a genuine Vermeer" depends on human conventions, on our social practices of attributing authorship, provenance, and artistic value. These are facts about us, not about the paint on the canvas. Two physically identical paintings really do have different properties (market value, art-historical significance), but those properties are socially constructed.

Consciousness is not socially constructed. Whether a brain produces subjective experience does not depend on human conventions, social practices, or anyone's opinion. It depends on what that brain is doing, physically, right now. The question "Is this brain conscious?" is a question about the brain's present physical state, not about its history or provenance. The Vermeer analogy fails precisely because it confuses a socially constructed property (authenticity) with a physical one (consciousness).

That said, I acknowledge that this is a genuine philosophical disagreement, not a settled question. Reasonable people can disagree about whether causal history matters for identity. What I want to argue is that the burden of proof lies with those who claim it does, because they are claiming that something beyond the present physical state of the brain affects consciousness, and we have no evidence that anything beyond the present physical state affects consciousness.

Objection Two: The Hard Problem and the Explanatory Gap (Chalmers's Challenge)

David Chalmers, arguably the most important living philosopher of mind, formulated what he calls "the hard problem of consciousness." The easy problems of consciousness (explaining how the brain processes information, controls behavior, integrates data) are hard in practice but conceptually straightforward. They are problems about what the brain does. The hard problem is about why the brain's doing is accompanied by subjective experience at all. Why does there feel like something to be a brain processing information?

The hard problem is a fundamentally different kind of challenge, not a gap in our knowledge that will eventually be filled. It is, Chalmers argues, a fundamentally different kind of problem. Physical explanations describe structure and function: what things are made of and how they work. But subjective experience is not a structure or a function. It is what it feels like to have a structure that functions in a certain way. And it is not clear that any amount of structural and functional description can explain why there is something it feels like.

This matters for Ixperiencit Theory because my theory rests heavily on the claim that physical organization produces consciousness. If Chalmers is right that there is an unbridgeable explanatory gap between structure/functioning and consciousness, then my premises may be incomplete. There may be more to consciousness than the physical pattern, some additional ingredient, some psychophysical law, that determines when and how subjective experience arises from physical processes.

My response: I take the hard problem seriously. It is a genuine puzzle, and anyone who claims to have solved it is almost certainly wrong. But I want to distinguish between two versions of the hard problem, because they have very different implications for Ixperiencit Theory.

Version 1: The Explanatory Gap. We don't understand how physical organization produces consciousness. This is true, and it may remain true for a long time. But it doesn't undermine our three premises. We didn't understand how gravity worked for centuries after Newton quantified it. The explanatory gap didn't make gravity less real or less predictable. Similarly, not understanding the mechanism by which brains produce consciousness doesn't change the strong evidence that they do, or that identical brains produce identical consciousness.

Version 2: The Ontological Gap. Consciousness might be a fundamental feature of reality, not reducible to physical organization at all. This is Chalmers's more radical suggestion, and it includes possibilities like panpsychism (consciousness is everywhere, in some form) or property dualism (consciousness is a non-physical property that accompanies certain physical arrangements).

If the ontological gap version is correct, it could go one of two ways for Ixperiencit Theory:

The friendly version: Consciousness accompanies the physical pattern according to some psychophysical law. Identical organization always produces identical consciousness, because the psychophysical law is consistent. In this case, all three of our premises still hold. The additional ingredient (the psychophysical law) doesn't change the conclusion; it just adds a deeper explanation of why the premises are true.

The unfriendly version: Consciousness is determined by something other than, or in addition to, the physical pattern. Two identical brains could produce different conscious experiences, or no conscious experience at all (the p-zombie scenario). In this case, our premises would be wrong, and the theory would fail.

The unfriendly version would be devastating, but there is no evidence for it. Every piece of evidence we have points to a tight, consistent relationship between brain structure/functioning and consciousness. Alter the brain, and you alter consciousness. Restore the brain, and you restore consciousness. Two similar brains produce two similar consciousnesses. The unfriendly version asks us to believe that this remarkably consistent relationship could fail — that an identical brain could produce different consciousness — despite having zero cases where this has been observed.

I cannot prove the unfriendly version is impossible. But I can say that the evidence strongly favors the friendly version, and that Ixperiencit Theory's conclusions hold under either the friendly version or the simple premise that the physical pattern is all there is.

Objection Three: The Measure Problem and the Probability Question

Here is an objection that comes not from philosophy but from physics and mathematics. Even if we grant that identical physical organization produces identical consciousness, a critical question remains: **How likely is it that your pattern will actually be reproduced?**

This is what physicists call a "measure problem," a problem about how to assign probabilities in a potentially infinite space. And it is a genuine challenge for the theory.

The argument goes like this. The human brain contains roughly 86 billion neurons14, each with an average of 7,000 synaptic connections to other neurons. Together, the precise strengths, positions, and properties of all these connections, along with the chemical milieu, the genetic expression, the ongoing electrical dynamics, constitute a state of staggering complexity. And the number of possible brain states is astronomically large, far larger than the number of atoms in the observable universe.

For your specific brain state to be reproduced by chance, the universe would need to randomly assemble the right configuration from the unimaginably vast space of possible configurations. Even in an infinite universe, the frequency of your exact configuration might be so sparse that it is effectively never encountered.

My response: This objection conflates two different claims that Ixperiencit Theory makes, and I want to carefully separate them.

Claim 1 (Logical): If a brain with your physical organization exists, it produces your consciousness. This is a claim about the relationship between structure/functioning and consciousness. It does not depend on probability or cosmology. It is true whether or not any such brain will ever actually exist again.

Claim 2 (Cosmological): Such brains will exist again, because the universe is large enough and long-lasting enough to guarantee it. This is a claim about the physical universe, and it involves assumptions about cosmology, probability, and the large-scale structure of reality.

Claim 1 is the core of the theory, and it stands regardless of what happens with Claim 2. Even if the universe were going to end tomorrow, the principle that your consciousness could be reproduced would still be true. The principle is about the nature of consciousness, not about the size of the universe.

Claim 2 is more speculative, and I want to be honest about that. It depends on cosmological assumptions that are themselves uncertain:

- **Is the universe infinite?** Current evidence suggests it might be, but we don't know for sure. If the universe is finite, the probability calculus changes.
- **Is the universe ergodic?** An ergodic system is one that, given enough time, explores all possible states. If the universe is

ergodic, every possible brain state occurs eventually. But we don't know if the universe is ergodic.

- **What is the relevant timescale?** Even if your brain state will eventually recur, "eventually" might mean after a time span so vast that it dwarfs the age of the universe by factors of 10^10^10 or more.
- **Does the Boltzmann brain problem apply?** A "Boltzmann brain" is a brain that forms spontaneously from random fluctuations. Some physicists argue that in an infinite universe, most conscious experiences would be produced by Boltzmann brains: random, momentary fluctuations rather than evolved, situated brains. If so, your future "instances" might be bizarre, fleeting, contextless flickers of consciousness rather than full lives. This is a serious concern.

I do not want to sweep these difficulties under the rug. The cosmological claim is the most speculative part of the theory, and future physics will need to address it. What I want to emphasize is that the logical claim, Claim 1, stands on its own, and it is the more important claim. It changes how we think about consciousness, identity, and death, regardless of the specific predictions about how often your consciousness will be reproduced.

Moreover, the cosmological claim does not depend solely on random chance. It also includes the possibility of deliberate reproduction — conscious beings with advanced technology choosing to reproduce specific brain states. This is far more likely than random reproduction, and it does not face the same measure-problem difficulties.

Objection Four: The "So What?" Objection

This is perhaps the most emotionally resonant objection. Grant every premise of the theory. Grant that your ixperiencitness could, in principle, be reproduced. The objection is: **So what?**

Critics argue: the "me" that will exist in the future doesn't help the "me" that exists now. You will still die. You will still lose your relationships, your projects, your life. The knowledge that some future entity will have your ixperiencitness is abstract consolation at best. What good is a theory of immortality if you still lose everything you care about?

My response: This objection comes from a genuine place of human concern, but it rests on a misunderstanding of what the theory actually implies.

The future instance of your ixperiencitness is not a stranger who happens to share your properties. It is you — the same conscious experience, the same ixperiencitness, the same subjective reality. And the space of possible paths your ixperiencitness could take is not limited to alien, unfamiliar futures. It includes paths where you experience reunion with people you love, paths where you experience states of bliss that match any religious vision of heaven, and paths that go far beyond anything religion has conceived, including the possibility of true empathy, of actually experiencing the conscious lives of others from the inside.

Yes, there is a real loss in death. The specific relationships, projects, and ongoing narratives of this life end. That is genuinely sad. But what superimmortality changes is the scope of what comes next, far beyond the finality of death. The claim is not merely that death is not the last word. It is that the next words may be richer and deeper than anything in this life.

The loss of this life is real. But what lies ahead is not a diminished echo of what you had. It is the full space of possible conscious experience, including experiences that encompass and surpass everything you value now.

What These Objections Don't Do

None of these objections proves Ixperiencit Theory wrong. What they do is identify the points where the theory makes claims that go beyond what current science can definitively prove. These are the pressure points, the places where future evidence and argument will either strengthen or weaken the theory.

The causal-continuity objection challenges whether history matters for identity. Meanwhile, the hard problem challenges whether the physical pattern is the complete explanation for consciousness. The measure problem asks whether the universe will actually produce future instances of your consciousness. And the "so what?" objection challenges whether the theory's implications matter on a personal level, though as Chapter 27 shows, they matter deeply.

CHAPTER 19

What Would Falsify This Theory?

A theory that cannot be proven wrong is not a scientific theory. It is a belief. And I have no interest in beliefs.

So let me state clearly what evidence would falsify Ixperiencit Theory, what discoveries would force me to abandon or fundamentally revise the theory. Any theory worth taking seriously must be willing to specify the conditions of its own failure, and this one is no exception.

Falsification Condition 1: Consciousness Does Not Depend on Structure and Functioning

If it were demonstrated that consciousness can arise without physical structure and functioning (that a soul, a spirit, or some non-physical substance can produce conscious experience independently of any brain or brain-like system) then Premise One fails, and the entire theory collapses.

What would this look like? It would look like credible, reproducible evidence of consciousness persisting after the complete destruction of the brain. Not near-death experiences (which occur while the brain is still functioning, however impaired). Not anecdotal reports of ghosts or communication with the dead (which have never been confirmed under controlled conditions). But something like: a person whose brain is verifiably, completely non-functional, yet who demonstrably communicates novel, verifiable information.

After more than a century of parapsychological research, no such evidence exists. Every purported case of consciousness without a brain has been explained by brain activity that was not yet fully stopped, by cold reading, by cognitive biases, or by fraud. The evidence consistently points in one direction: no brain activity, no consciousness.

But I state the condition openly: show me consciousness without a brain, and Premise One falls.

Falsification Condition 2: Identical Structure and Functioning Produces Different Consciousness

If it were demonstrated that two physically identical brains (same structure, same functioning, down to whatever level of detail turns out to be relevant) produce different conscious experiences, then Premise Three fails.

What would this look like? In practice, this is extraordinarily difficult to test, because we cannot yet build identical brains or measure consciousness directly. But here is a thought experiment that illustrates what would count.

Suppose we develop technology to build two brains that are identical at the relevant level of description (whatever that turns out to be: molecular, synaptic, connectomic). Suppose both brains are placed in identical environments and given identical sensory inputs. If the two brains reported different experiences — if one said "I see red" and the other said "I see blue" in identical circumstances — this would falsify Premise Three.

More realistically, if we found that identical twins with extremely similar brain structure systematically reported different qualia for identical stimuli, not just different preferences or associations, but genuinely different raw experiences, this would be suggestive (though not conclusive, since identical twins do not have identical brain structure at the relevant level of detail).

The key prediction of Ixperiencit Theory is that as brain structure converges, conscious experience converges. If we found cases where brain structure was nearly identical but experience was radically different, the theory would be in serious trouble.

Falsification Condition 3: History Matters Independently of Present State

If it were demonstrated that the causal history of a brain (how it came to be in its current state) affects consciousness independently of the brain's current physical organization, then the core logic of the theory breaks down.

What would this look like? Suppose we could build two brains in identical present states, but with different histories (one arrived at that state through normal biological development, the other was assembled artificially). If the two brains, despite being physically identical in the present moment, produced demonstrably different conscious experiences, this would show that history matters, that there is something about how a brain got to its current state, beyond the current state itself, that affects consciousness.

This would be evidence for something like a soul, an extra factor beyond physics, that tracks the history of a body and contributes to its consciousness. Finding such a factor would be one of the most important discoveries in the history of science. It would also falsify Ixperiencit Theory.

I should note that this is an empirical question, not a logical one. I believe the evidence strongly suggests that present state is all that matters. But I could be wrong, and I specify this as a falsification condition precisely because I want the theory to be testable rather than unfalsifiable.

Falsification Condition 4: Consciousness Is Not Substrate Independent

If it were demonstrated that consciousness depends on the specific material substrate (that biological neurons produce consciousness but identical structures made of silicon, or some other material, do not) then the substrate-independence claim fails.

This would not necessarily destroy the entire theory (your ixperiencitness could still be reproduced by other biological brains), but it would significantly narrow its scope. And it would raise deep questions about why the specific material matters, since the laws of physics treat identical structures identically regardless of what they're made of.

What would this look like? If we built an artificial system that perfectly replicated the organization of a biological brain — same connectivity, same dynamics, same information processing — and it passed every behavioral test for consciousness but could be shown (by some future measurement technique) to lack subjective experience, this would challenge substrate independence.

This is related to the "Chinese Room" argument by philosopher John Searle, who argued that a computer simulating a brain is no more conscious than a person simulating understanding of Chinese by following rules. Ixperiencit Theory disagrees with Searle, predicting that a system with the right organization will be conscious regardless of substrate. But this is an empirical prediction that could, in principle, be tested.

Falsification Condition 5: The Universe Cannot Reproduce Your Brain State

This is a weaker form of falsification — it would not destroy the logical claims of the theory but would undermine the cosmological claim that your consciousness will actually be reproduced.

If it were demonstrated that the universe is finite, closed, and heading toward heat death in a way that makes the reproduction of specific brain states physically impossible (not just unlikely, but impossible) — then the prediction that you will live again becomes a prediction about what advanced civilizations might achieve, not a prediction about what the universe guarantees.

I want to be honest: this is the part of the theory most vulnerable to future physics. We do not know whether the universe is infinite or whether it is ergodic. Nor do we know the precise probability of any given brain state recurring naturally. The cosmological claim rests on assumptions that are themselves uncertain.

But note: even if this condition were met, the principle of Ixperiencit Theory would still hold. Even in a finite universe heading for heat death, it would still be true that a brain with your physical organization would produce your consciousness. The claim "identical S&F produces identical consciousness" does not depend on the universe being large enough to produce duplicates. It is a claim about the nature of consciousness, not about the size of the cosmos.

What Cannot Falsify the Theory

It is equally important to state what does not count as falsification:

"It feels wrong." Intuitions about identity are not evidence. As we discussed in Chapter 3, our intuitions about consciousness were shaped by evolution for survival, not for truth. Saying "a replica wouldn't really be me" is like saying "the Earth can't really be moving" — it expresses a feeling, not an argument.

"We can't currently test it." Many scientific theories were untestable when first proposed. The Higgs boson was predicted in 1964 and not confirmed until 2012, nearly five decades later. A theory being currently untestable does not make it unscientific, as long as it specifies what would test it — which I have done above.

"It sounds like religion." Structural similarity to religious claims is not evidence against a scientific theory. Religions claimed the universe had a beginning long before the Big Bang theory confirmed it. The fact that Ixperiencit Theory reaches an optimistic conclusion about death does not make it religious, any more than the fact that modern medicine reaches optimistic conclusions about disease makes medicine religious.

"Other philosophers disagree." Philosophy of mind is a field of vigorous debate, and there is no consensus on the nature of consciousness. The fact that some philosophers (Parfit, Chalmers, Searle) would object to aspects of this theory does not falsify it. What matters is the quality of the arguments, not the authority of the arguers. I have engaged with the strongest objections above, and I believe the theory withstands them.

The Honest Position

Here is where I stand: Ixperiencit Theory is built on well-supported scientific premises and reaches surprising conclusions. It makes specific, testable predictions about the relationship between brain structure and consciousness. Some of these predictions are testable with current technology (or technology we can foresee). Others will require scientific advances we cannot yet imagine.

I have specified five conditions that would falsify the theory. I invite scientists and philosophers to test them. If any of these conditions is met, I will revise or abandon the theory — because I am interested in truth, not in being right.

But until one of these conditions is met, the argument's logic is straightforward. If that is true (and the evidence strongly suggests it is) then patterns, by their very nature, can exist again.

CHAPTER 20

The Mathematics of Recurrence

So far in this book, I have made two different kinds of claims, and I have not always been careful to distinguish between them. It is time to be precise, because the distinction matters enormously.

The logical claim: If a physical system has the same organization as your brain, it produces the same consciousness. This is a claim about the nature of consciousness. It follows from our three premises and does not depend on cosmology, probability, or the size of the universe.

The cosmological claim: Such physical systems will actually exist — elsewhere in the universe, or in the far future, or both. This is a claim about the physical world, and it requires its own justification.

This chapter is about the cosmological claim. How confident can we be that your consciousness will actually be reproduced? What does the math say? And where does the math run into trouble?

The Numbers

Let's start with the raw numbers, because they are instructive.

The human brain contains approximately 86 billion neurons. Each neuron has, on average, about 7,000 synaptic connections to other neurons, giving roughly 600 trillion synapses. Each synapse has a "strength" that can vary continuously, and each neuron has a firing threshold that depends on its chemical environment.

How much information does this represent? Estimates vary widely, but a reasonable lower bound is that the brain's state can be described by somewhere between 10^14 and 10^16 bits of information, somewhere between 100 terabytes and 10 petabytes.15 (The exact number depends on how finely we need to specify each synapse, each neurotransmitter concentration, each quantum state; a question we'll return to.)

The space of possible brain states is therefore approximately 2^(10^15) — two raised to the power of a quadrillion. This number is so large that writing it out in standard notation would require more digits than there are atoms in the observable universe. It is, for all practical purposes, infinite.

This upper bound vastly overstates the number of biologically realizable brain states, since neural configurations are heavily constrained by physics, anatomy, and the developmental processes that build brains. The actual space of possible human brains is far smaller — though still unimaginably large.

Your specific brain state occupies one point in this unimaginably vast space. For the universe to reproduce your brain state by chance, it would need to randomly sample from this space and hit your exact point.

The Vast Universe

Now consider the size of the playing field.

The observable universe contains roughly 10^80 atoms and has existed for about 13.8 billion years. But "observable" is a limitation of our perspective, not a limitation of reality. Many cosmological models suggest the total universe is far larger than the observable portion, perhaps infinitely larger.

If the universe is infinite in spatial extent, it contains infinitely many atoms. In an infinite universe, any configuration of matter that has a nonzero probability of occurring does occur — and it occurs infinitely many times. This is not a speculation; it is a mathematical theorem about infinite sample spaces. If the probability of your brain state occurring in any given region of space is epsilon (some tiny positive number), then the expected number of occurrences in an infinite universe is infinity × epsilon = infinity.

This is the basis for the **cosmic recurrence argument**: in an infinite universe, every possible brain state exists somewhere, including yours. You are a point in the vast space of consciousness that is instantiated infinitely many times.

But Is the Universe Actually Infinite?

Here is where intellectual honesty requires us to slow down.

We don't know if the universe is spatially infinite. Measurements from the Planck satellite mission (2018) are consistent with both a finite and an infinite universe. The curvature of spacetime appears to be flat (or very close to flat), which is consistent with infinity, but a very large finite universe with slight curvature would look the same to our measurements.

We also don't know whether the universe will last forever. The current best understanding suggests it will — that the universe will expand forever, cooling and diluting but never contracting or ending. But cosmology is a young science, and our understanding of the far future of the universe is necessarily speculative.

And even if the universe is infinite and eternal, there is the question of whether it is **ergodic**, that is, whether it actually explores all possible configurations of matter over time. A universe could be infinite and eternal but still be constrained by physical laws to only visit certain configurations, never reaching others.

The Boltzmann Brain Problem

There is an even more troubling challenge for the cosmological claim, and it comes from statistical mechanics. Ludwig Boltzmann, the 19th-century physicist who founded statistical thermodynamics, realized that in a universe in thermal equilibrium, random fluctuations will occasionally produce low-entropy states16 — including, very rarely, states that look like brains.

A **Boltzmann brain** is a brain that forms spontaneously from random fluctuations in an otherwise featureless thermal soup. It has all the structure and functioning of a real brain (including memories, beliefs, and conscious experience) but it exists only for a fleeting moment before dissolving back into entropy.

The Boltzmann brain problem is this: if you're calculating the probability of your brain state being reproduced somewhere in the cosmos, most of the instances might be Boltzmann brains — momentary fluctuations that pop into existence with your memories and then immediately dissolve. These would be instances of your consciousness, according to the theory, but they would not be lives. They would be single awaremoments, or at most very brief sequences of awaremoments, surrounded by non-existence.

This is genuinely unsettling, and I want to address it honestly rather than hand-wave it away.

There are several responses:

First, the Boltzmann brain calculation depends on assuming that the universe reaches thermal equilibrium and stays there for extremely long periods. If the universe never reaches true equilibrium, if new structures continue to form, if cosmic inflation produces new regions of complex matter, then the Boltzmann brain scenario may not apply. Indeed, many cosmologists treat the prediction of Boltzmann brains not as a genuine forecast but as a reductio ad absurdum — evidence that the cosmological model producing the prediction must be wrong, rather than evidence that random brains will actually form.

Second, even if Boltzmann brains exist, they don't undermine the logical claim. The principle that identical physical organization produces identical consciousness is true whether the system is a Boltzmann brain or a biologically evolved brain. What changes is the character of the future instances — whether they are full lives or momentary flickers.

Third, and most importantly, the Boltzmann brain scenario is not the only way your consciousness could be reproduced. Deliberate reproduction by advanced civilizations is far more likely to produce full, sustained, enhanced conscious experiences, and it does not depend on random fluctuations or thermal equilibrium.

Deliberate vs. Random Reproduction

This brings us to a key distinction that changes the probability calculus entirely. Note that we are shifting here from mathematics and physics, where the reasoning is rigorous, to predictions about the behavior of future civilizations, where we are necessarily on less certain ground.

Random reproduction depends on the universe blindly assembling your brain state through chance. The probability of this is indeed astronomically small for any given region and time period, though it may be nonzero in an infinite universe.

Deliberate reproduction depends on intelligent beings with advanced technology choosing to create or recreate specific brain states. This changes everything, because:

1. **Intelligence is a probability amplifier.** A civilization that understands the science of consciousness and has the technology to create specific brain states can produce any consciousness it wants, without waiting for random chance.

2. **Technology tends to advance.** The history of human technology is a history of turning the impossible into the routine. Five hundred years ago, recording a song was impossible. One hundred years ago, recording it at high fidelity was cutting-edge. Today, anyone with a smartphone can do it. The arc of technology bends toward greater control over matter.

3. **The motivation exists.** If superimmortality is understood and valued, there is strong motivation for advanced civilizations to reproduce and enhance conscious experiences. This is not idle speculation; it is the natural extension of values that humans already hold: the desire to preserve life, to reduce suffering, to understand consciousness.

4. **Multiple independent civilizations are likely.** The universe almost certainly contains other intelligent civilizations (the Drake equation, proposed by astronomer Frank Drake in 1961, yields estimates ranging from near-zero to millions of civilizations depending on the inputs17). Each is an independent source of potential deliberate reproduction.

The point is that the cosmological claim does not rest solely on the dubious proposition that your brain state will randomly recur. It also rests on the more plausible proposition that conscious beings with advanced technology will deliberately produce specific brain states, including states that reproduce your ixperiencitness.

What the Math Actually Tells Us

Let me summarize the mathematical situation honestly:

We can say with confidence:

- The space of possible brain states is unimaginably large.
- Your specific brain state is one point in that space.
- If the universe is infinite, every brain state with nonzero probability occurs infinitely many times.
- If the universe is finite, the probability of random recurrence depends on details we don't yet know.

We cannot say with confidence:

- Whether the universe is actually infinite.
- Whether the universe is ergodic.
- How precisely a brain must be replicated to reproduce a given ixperiencitness. (If perfect atomic-level fidelity is required, recurrence is far less likely than if approximate neural-level fidelity suffices.)
- The relative frequency of Boltzmann brains versus evolved brains among future instances.
- What advanced civilizations will actually choose to do with their technology.

The takeaway: The cosmological claim is not proven. It is a reasonable extrapolation from what we know about physics and cosmology, but it depends on assumptions that are themselves uncertain. This is where the theory is most speculative and most in need of future scientific input.

But the logical claim, that identical physical organization produces identical consciousness, does not depend on cosmology at all. It is a claim about the nature of consciousness, and it stands regardless of the size, age, or fate of the universe.

The mathematics of recurrence should not be oversold. What the math gives us is not certainty but possibility: possibility grounded in real physics, real cosmology, and the real tendency of the universe to produce complex structures. Whether that possibility is realized is a question for future science, future technology, and future civilizations.

What the math does not give us is any reason to think that death is permanent. Even in the most pessimistic cosmological scenario, the logical claim holds: your consciousness could be reproduced. Whether it will be is a separate question, and it is a question worth working to answer.

PART FIVE: WHY YOU ARE IMMORTAL

CHAPTER 21

Why Continuity Doesn't Matter

Here is one of the most common objections to the line of thinking in this book: "But there's a gap! The original dies, and the replica is somewhere else. There's no continuity. Without continuity, it's not really me."

This objection feels powerful, but it dissolves under scrutiny.

First, let's distinguish between two concepts that people often confuse: **continuity** and **continuousness**.

Continuity means sameness between two states. Your brain at noon is continuous with your brain at 12:01 if the structure and functioning are similar enough to produce the same identity.

Continuousness means uninterrupted existence. Your consciousness is continuous if there is no gap or break in it.

Most people assume that continuousness is necessary for identity, that you need an unbroken chain of conscious moments to remain "you." But this is clearly wrong, and you can prove it to yourself tonight.

The Sleep Gap

When you go to sleep, your consciousness is interrupted. During deep sleep, there is no conscious experience (or at best, very minimal, disconnected experience that you don't remember). You are, for all practical purposes, unconscious. Then you wake up, and you are you again.

What happened during the gap? Nothing, as far as consciousness is concerned. The structure of your brain was maintained (roughly), and when it resumed functioning in the right way, your consciousness returned. The gap didn't make you a different person.

Under general anesthesia, as we discussed, the gap is even more dramatic. Your consciousness is completely obliterated, more thoroughly than in sleep. Yet when you wake up, you are still you.

So continuousness, an unbroken stream of consciousness, is clearly not required for identity. What matters is the structure and functioning when consciousness resumes. If the structure and functioning are right, the consciousness is right, regardless of what happened in between.

The Billion-Year Gap

Now apply this to death.

If your brain's structure could be perfectly reproduced after death (whether a minute later, a year later, or a billion years later) and that reproduced brain began functioning, the consciousness it produced would be you. The gap between your death and the reproduction would be like the gap of sleep or anesthesia: irrelevant from the perspective of conscious experience, because there is no experience of the gap.

Imagine this. You die today. A billion years from now, by chance or by design, a brain comes into existence with the same structure and functioning yours had at some point during your life. Consciousness begins in that brain. That consciousness has your memories, your personality, your ixperiencitness.

What does this feel like from the inside?

It feels like waking up. One moment you were here, reading this book. The next moment, instantly, with no sensation of intervening time, you are there, a billion years in the future, in a different body, in a different world, with no memory of the gap (because there was nothing to remember). You simply continue where your awarepath left off, or perhaps where it left off at some earlier point, depending on the structure of the brain that was reproduced.

This is not science fiction. It follows directly from the three premises we established in Part One, combined with the insight, demonstrated in the previous section, that gaps in consciousness don't affect identity. If those premises hold, then temporal gaps of any duration are irrelevant.

The Continuity of the Pattern

The science of superimmortality replaces the traditional notion of bodily continuity with the notion of **continuity of the pattern**. These are very different things.

Bodily continuity means: the same physical body, made of the same matter, existing continuously through time. This is what most people think identity requires. But as we have seen, it is neither necessary (because atoms are replaced) nor sufficient (because a body can be radically restructured while maintaining physical continuity, as in the anesthesia extension argument).

Continuity of the pattern means: the relevant physical organization (the brain's arrangement and operation) is preserved or reproduced, whether in the same body or a different one, whether continuously or with gaps.

What matters is the pattern, and the pattern can exist again.

CHAPTER 22

You Have Already Survived Death — Many Times

Here is something most people have never considered: by the logic of this book, you have already survived multiple "deaths," and you didn't even notice.

The You of Age Five Is Gone

Think back to when you were five years old. Try to remember what it was like. Maybe you can summon a few images: a house, a parent's face, a toy. But the actual experience of being five (the way the world felt, the way your thoughts worked, the way you processed emotions) is almost entirely lost to you. Not just the memories. The way of being conscious.

The five-year-old you had a fundamentally different brain — a brain still under heavy construction, with far more synaptic connections than your adult brain but far less refined wiring (your brain didn't finish developing until your mid-twenties). Different cognitive abilities (you couldn't do algebra, understand irony, or think abstractly), different emotional patterns, and different personality traits. And, of course, different atoms; every single atom in the five-year-old's brain has long since been replaced.

In what sense is that five-year-old you?

Not in the sense of having the same atoms (they're all gone). Not in the sense of having the same brain structure (it was dramatically different). Not even in the sense of having the same ixperiencitness (the five-year-old's experience of the world was qualitatively different from yours today).

The five-year-old you was, in the language of Ixperiencit Theory, a different point on the ixperiencitness continuum. Connected to the current you by a continuous chain of gradual changes, but significantly different in the actual quality of conscious experience.

The Continuum of Self

This reveals something important about identity: **you have been slowly becoming a different person your entire life.**

At every stage (infancy, childhood, adolescence, young adulthood, middle age, old age) your brain's organization changes. Your ixperiencitness shifts. Not suddenly, not dramatically (usually), but continuously. The you of 20 is different from the you of 40, which is different from the you of 60. Each is recognizably connected to the others, but none is identical.

This is not a crisis for the concept of identity. It is what identity actually looks like, once you let go of the folk-psychology notion that "you" are a single, unchanging thing. Identity is not a point. It is a trajectory, a path through the space of possible ixperiencitnesses, meandering from one configuration to another over the course of a lifetime.

Your awarepath is not a straight line. It curves, it evolves, it transforms. The person at the beginning of the path shares ixperiencitness with the person at the end — but they are not the same person in the way we colloquially mean "same." They are connected points on a continuum.

What This Means for Death

The point of all this is to dissolve the sharp line between "alive" and "dead," and between "you" and "not-you."

You are not a static thing that either exists or doesn't. You are a process, a continuous, evolving pattern of structure and functioning that changes every moment, sometimes gradually and sometimes dramatically. The process has interruptions (sleep, anesthesia). It has slow transformations (aging, learning, brain plasticity). It has sudden changes (injury, illness, transformative experiences).

Death is the permanent cessation of this process in one particular body. Consciousness ceases because the brain stops functioning in the right way.

And resurrection, the reproduction of your ixperiencitness by another body, is a more extreme version of what happens every morning when you wake up: consciousness resumes because a brain is functioning in the right way.

You have survived the transition from infancy to adulthood, a transformation so complete that the infant and the adult share almost nothing in terms of specific brain structure. Every night, you survive a cessation of consciousness. And if you have had surgery, you have survived the total elimination of consciousness under anesthesia.

If you survived all of these, what makes death different? Only one thing: the gap is longer, and the body that produces the next consciousness is a different one. But we have already established that the body doesn't matter — the pattern matters. And we have already established that the length of the gap doesn't matter — because you don't experience the gap.

Sleep and anesthesia are not death — the differences are real. But the experiential structure is shared: in all three cases, consciousness ceases, a gap occurs that you do not experience, and consciousness resumes in a brain that is not identical to the one that existed before. The differences are in degree, not in kind.

If you accept that you survive the nightly cessation of consciousness despite waking up with a slightly different brain, then the question about death becomes not whether you can survive a cessation of consciousness, but how large a change you can survive. And we have already seen, in the case of the five-year-old, that the answer is: a very large one indeed.

You have been navigating the relevant features of this transition your entire life. The precedent is there. The question is only whether you are willing to follow it to its conclusion.

CHAPTER 23

Superimmortality: Beyond Ordinary Survival

When most people think of immortality, they imagine simply continuing to live forever, never dying, never ending. Eternal youth, or at least eternal existence, in a single body. But this is actually a very limited vision of what is possible.

The claim is extraordinary. What I want to show is that it follows directly from premises you have already accepted, and that it is more detailed, more interesting, and more grounded than the word "immortality" might suggest.

Ixperiencit Theory reveals something far greater than simple immortality. I call it **superimmortality**.

The word "immortality" is actually inadequate, because technically no one's body will exist forever without dying. Eventually the sun will engulf the Earth, and the universe will undergo heat death. No single physical system lasts forever. If immortality requires one body persisting for eternity, then immortality is impossible.

But superimmortality doesn't depend on any single physical system. Superimmortality is the recognition that your conscious experience, your ixperiencitness, will be produced again and again, by different physical systems, across all of space and time. It is not about one body surviving forever. It is about the pattern of your consciousness being reproduced, in various forms, endlessly.

The Distinction Between Mortal and Supermortal

Consider a new word: **supermortal**. Being mortal means dying once and being gone forever. Being supermortal means dying many times, in many bodies, across many times and places, but also coming back to life many times, in many bodies, across many times and places.

Supermortality is the natural condition of any conscious being, given the premises we have established. If the universe is vast enough and long-lived enough, the structures and functionings that produce your consciousness will arise again and again. You will die again and again, but you will also live again and again. Death is not an ending; it is a transition between instances.

Superimmortality goes further. It means not just repeated existence, but enhanced existence. Not just the same life over and over, but better lives, richer lives, more varied lives. The premises established in Parts One through Three logically entail that your consciousness is reproducible; the forms described below are what the theory predicts when we apply that principle across the full scope of space, time, and technology. Superimmortality encompasses:

Multiple simultaneous existences. Right now, there may be physical systems elsewhere in the universe that produce consciousness identical to yours. You are experiencing being all of them, because the experience is identical to yours. You don't know you are experiencing them, because your current brain doesn't have the neural connections to represent that knowledge. But the experience is there.

Enhanced existences. Your ixperiencitness doesn't have to remain static. A brain structured similarly to yours but enhanced (more intelligent, more perceptive, more capable) could still produce a consciousness that is recognizably you, but better. You would experience being a genius, an artist of remarkable ability, a being of deep understanding. These enhanced versions of you are still you, because they share your ixperiencitness (perhaps enhanced, but continuous with yours).

Existences across time. There is no requirement that your consciousness be produced only in the present. If, billions of years from now, a physical system arises that produces your brain's structure and functioning, it produces your consciousness. From your perspective, you will experience that future existence just as you experience the present one.

Existences in radically different bodies. Your consciousness is produced by the brain's physical organization, but brains don't have to be biological. A sufficiently advanced artificial system (electronic, quantum, or based on physics we haven't yet discovered) could reproduce the relevant organization of your brain. You could experience existence in a machine, in a virtual world, in forms we cannot yet imagine.

Controlled and designed experiences. With sufficient technology, a brain's inputs could be precisely controlled to produce any desired conscious experience. Superimmortality includes the possibility of conscious existences that are specifically designed to be fulfilling, joyful, enlightening, or adventurous.

This is not eternal life in a single body. This is eternal existence across all possible bodies, all possible times, all possible worlds. It is existence without limit.

Few visions of the afterlife have been this expansive. And unlike religious promises, this one is grounded in what we already know about how consciousness works, though, as we explored in Part Four, some aspects remain genuinely uncertain.

But what forms does this immortality take, and how certain is each one? Ixperiencit Theory maps out five distinct forms, each with different characteristics and degrees of certainty.

Form 1: Natural Immortality

This is the most fundamental form, because it depends only on our basic premises and the nature of the universe: no technology, no deliberate action, no one intending to reproduce your consciousness.

Somewhere in the incomprehensible vastness of space and time, a brain will form — by chance or by the workings of physics we do not yet understand — with your pattern. It will open its eyes and experience being you, with no memory of this life but with the same quality of consciousness. This is natural recurrence: the universe's brute-force guarantee.

If consciousness is produced by physical organization, and if the universe is vast enough and long-lived enough, then the physical patterns that produce your consciousness will arise again — somewhere, sometime. The universe, simply by being big enough and old enough, produces brains similar to yours as a matter of statistical inevitability.

I should be transparent about the uncertainty here. This form of immortality depends on cosmological assumptions we cannot yet confirm: that the universe is infinite or sufficiently large, that it is ergodic (explores all possible configurations), and that the relevant brain states have nonzero probability of arising naturally. These are open questions in physics. Natural immortality is the form with the strongest philosophical foundation but the most uncertain physical grounding. (We examine this uncertainty carefully in the chapter on the Mathematics of Recurrence.)

Form 2: Enhancement Immortality

Enhancement immortality goes beyond mere reproduction. In this form, your consciousness is improved, not merely preserved.

Imagine waking one morning to find the world richer than it was the night before: colors more saturated, music more layered, ideas connecting in ways that once eluded you. You are sharper, deeper, more alive, yet still unmistakably yourself. This is enhancement immortality: you, continued and upgraded.

An enhanced version of you would have a brain structure similar to yours but upgraded in some way: higher intelligence, better memory, greater emotional capacity, new sensory abilities. The ixperiencitness would be recognizably yours, continuous with yours in the deep patterns that define your identity, but better. You, upgraded.

Think of it this way. The you of today is an enhanced version of the you before you learned to read. Before literacy, entire dimensions of thought were closed to you: abstract reasoning, written communication, the ability to absorb the accumulated knowledge of civilization. Acquiring that single skill transformed your cognitive architecture so profoundly that your pre-literate and post-literate selves are almost different kinds of minds. Yet you are still you — your ixperiencitness has evolved but maintained continuity. Enhancement immortality is an extension of this natural process of growth and development, beyond the limits of a single biological lifetime.

This is one of the most exciting aspects of superimmortality: the best is always yet to come.

Form 3: Multiplicity Immortality

This form is based on the existence of multiple simultaneous instances of your consciousness. If many physical systems exist at the same time, all producing your ixperiencitness, then the death of any one of them does not end your consciousness. The others continue.

Right now, on a planet orbiting a star you will never see, a brain may be producing an experience indistinguishable from yours, the same texture of thought, the same felt quality of being. Neither of you knows the other exists, yet the experience is one experience, lived twice. This is multiplicity immortality: consciousness insured by redundancy.

Multiplicity immortality does not require exotic physics. It follows directly from the premises: if many brains can produce the same ixperiencitness, and some of them exist simultaneously, then consciousness is distributed across multiple instances. Losing one instance is a loss, but it is not extinction.

Form 4: Technological Immortality

This form encompasses all the types of immortality that become possible with sufficiently advanced technology: brain scanning and digital reconstruction, uploading consciousness to artificial substrates, creating devices that produce specific conscious experiences by directly controlling brain-like structures.

Picture a future scientist scanning a preserved brain at the resolution of individual synapses, feeding the data into a system that rebuilds its pattern in a new substrate, and watching as the reconstruction opens its eyes with the same felt sense of being that the original once had. This is technological immortality: consciousness reproduced not by cosmic accident but by deliberate human effort.

Technological immortality overlaps with the transhumanist vision of the future, but goes further. Transhumanism typically focuses on keeping one body alive or transferring one consciousness to a new substrate. Technological immortality includes all of this, plus the deliberate creation of multiple, enhanced, and novel versions of your consciousness.

This is also the form that is most within our control. Every advance in neuroscience, every improvement in brain-scanning technology, every step toward understanding the minimal requirements for producing specific conscious experiences: all of these bring technological immortality closer.

Form 5: Fragmented Immortality

This form is both the most modest and the most immediately real. It is the immortality of fragments, pieces of your ixperiencitness that persist in the brains of other people.

A child laughs exactly the way her mother laughed, the same timing, the same bright surprise in the sound, and for an instant the mother's pattern of consciousness flickers back into the world, not whole, but real. This is fragmented immortality: pieces of you persisting in the minds you have touched.

Your children carry fragments of your neural architecture through genetics. Students carry fragments of your cognitive patterns through learning. And friends carry fragments of your emotional patterns through shared experience. None of these fragments, individually, constitutes full survival. But collectively, they represent a distribution of your ixperiencitness across other minds, a spreading, fragmenting, recombining pattern that persists long after any individual body dies.

Fragmented immortality requires no exotic physics, no advanced technology, and no cosmological assumptions. It is happening right now, through the ordinary mechanisms of genetics, culture, and human connection.

How These Forms Relate

These five forms are not mutually exclusive. They are different facets of the same underlying phenomenon: the reproduction of ixperiencitness across space and time. In practice, your superimmortality will likely involve all five forms in combination: natural recurrence of brain patterns, enhanced versions, simultaneous instances, technological reproduction, and the fragmented persistence of your patterns in other minds.

The forms vary in their certainty and immediacy. Fragmented immortality is happening now and is virtually certain. Technological immortality depends on scientific progress but is plausible. Enhancement and multiplicity immortality follow from the principles but depend on the right conditions. Natural immortality is philosophically grounded but cosmologically uncertain.

Together, they paint a picture of superimmortality not as a single promised afterlife but as a spectrum of possibilities (some certain, some probable, some speculative) all flowing from the same core insight about the relationship between consciousness and the physical patterns that produce it.

CHAPTER 24

Enhanced Versions of You

One of the most exciting implications of superimmortality is that your future existences need not be limited to what your current brain can produce. Enhancement is virtually certain, given enough time and the right conditions.

What Enhancement Means

An **enhaidentireplica** is a body whose brain structure and functioning is based on yours but improved in some way. The improvement could be in any dimension of conscious experience:

Cognitive enhancement. A brain with more neurons, more connections, or more efficient processing could produce a consciousness that thinks faster, remembers more, and understands more deeply — with your personality, your values, your way of seeing the world.

Sensory enhancement. Your current senses are limited by evolution to a narrow band of the electromagnetic spectrum and a narrow range of sound frequencies. An enhanced version of you could see infrared and ultraviolet, hear ultrasonic frequencies, feel the magnetic field of the Earth — or perceive through entirely new senses that have no analogue in current human experience.

Emotional enhancement. Deeper joy, more nuanced understanding, greater capacity for empathy — or the ability to experience complex emotional states that your current brain cannot produce, states for which there are no words in any language because no human has ever experienced them.

Memory and ability enhancement. Perfect recall, vast working memory, new creative and physical abilities: all are possible with the right brain organization.

The Continuity of Enhancement

An important question: if an enhanced version of you has a different brain structure, how can it still be you?

The answer lies in the continuum nature of ixperiencitness. Your ixperiencitness is not an all-or-nothing property. It can be partially shared between two consciousnesses. An enhaidentireplica would have the core patterns of your ixperiencitness (your personality, your cognitive style, your emotional character) while also having enhanced capabilities that you currently lack.

This is not fundamentally different from the way you change throughout your life. As we saw in Chapter 22, you have already survived transformations so radical that your earlier and later selves share almost nothing in terms of specific brain structure, yet your ixperiencitness has evolved while maintaining continuity.

Enhancement immortality is an extension of this process beyond the limits of a single biological brain. It is you, growing, learning, becoming more — across many lifetimes, in many bodies, with ever-expanding capabilities.

The Relationship Between Enhancement and Degradation

If enhancement is possible, so is its opposite: **degradation**. A brain with diminished organization would produce a consciousness that is recognizably yours but impoverished, less intelligent, less perceptive, less emotionally rich.

This is not a theoretical concern. It happens in real life, all the time. Aging degrades brain function. Diseases like Alzheimer's degrade brain structure. Strokes destroy specific brain regions. In each case, the person's consciousness is diminished: they lose memories, abilities, aspects of their personality.

From the perspective of ixperiencitness, degradation is the loss of aspects of yourself. The person with severe Alzheimer's still has some of your ixperiencitness — they are still, in some diminished sense, "you" — but they have lost much of what made your experience rich and full.

This cuts both ways. On one hand, it means that severe degradation is a kind of partial death, a loss of ixperiencitness even while the body continues to function. On the other hand, it means that even a degraded version of your consciousness is still partially you, still connected to your ixperiencitness on the continuum.

Enhancement and degradation are two directions on the same continuum. Understanding both is essential for a complete picture of superimmortality, because your future existences will not all be identical to your present one. Some will be enhanced. Some will be degraded. Most will be different in ways that include both enhancements and degradations in different dimensions of experience. The full scope of your superimmortality includes all of these variations.

CHAPTER 25

The Fragmented You: Consciousness Distributed Across Billions

There is one type of replica that deserves its own chapter, because it is happening right now, all around you, in a way you may never have considered: the **fidentireplica**.

A fidentireplica (a fragmented identical replica) is a physical system that produces part of your ixperiencitness. Not all of it. Just a fragment. A slice. A piece of what it's like to be you, instantiated in someone else's brain.

This is not science fiction. It is neuroscience.

You Are Already Fragmented

Consider your children, if you have them. They share roughly 50% of your DNA. But they also share something far more relevant to ixperiencitness: neural structures that are partially similar to yours. Not because they inherited your specific neurons, but because the same genetic blueprint that built your brain built theirs — and similar blueprints produce similar structures.

Your child's way of processing emotions, their cognitive style, their perceptual habits: some of these will overlap with yours, not because they learned them from you (though they learned some), but because their brains are built from a similar biological template. In the language of Ixperiencit Theory, your child is a fidentireplica of you: they produce a consciousness that shares fragments of your ixperiencitness.

But it goes far beyond genetics.

Cultural Fidentireplicas

When you teach someone something — really teach them, in a way that changes how they think — you alter their brain structure. You create new neural pathways, strengthen certain connections, shape the way they process information. Some of those changes make their brain slightly more similar to yours in specific ways.

A student who learns to think about a problem the way you think about it has, in a small way, acquired a fragment of your ixperiencitness. Not your memories, not your personality as a whole, but a specific cognitive pattern that you share.

Now multiply this across every person you have ever influenced. Every conversation that changed someone's perspective. Each argument that shifted someone's reasoning. Every joke that reshaped someone's sense of humor. Each act of kindness that altered someone's emotional life. Each of these is a tiny transfer of ixperiencitness, a fragment of your consciousness being reproduced, imperfectly and partially, in another brain.

This is fragmented immortality. It is happening constantly, invisibly, through the normal mechanisms of human social life. And it has been happening for as long as humans have communicated.

The Fidentireplica Spectrum

Not all fidentireplicas are created equal. The degree of fragmentation — the fraction of your ixperiencitness that is shared — varies enormously:

High-fidelity fidentireplicas: An identical twin is a very high-fidelity fidentireplica. Same genome (and therefore very similar brain structure), same early environment (often), similar cognitive and emotional patterns. Identical twins often report a deep sense of shared experience that goes beyond normal sibling closeness. This is not mystical; it is the natural result of similar brain structure producing similar ixperiencitness.

Medium-fidelity fidentireplicas: A child or a close student who has deeply internalized your way of thinking. They share significant aspects of your cognitive style, perhaps some emotional patterns, perhaps some characteristic ways of perceiving the world. But they also have much that is uniquely their own.

Low-fidelity fidentireplicas: Anyone who shares even one significant neural pattern with you. This includes anyone in your culture who processes information in broadly similar ways, anyone who shares your language (language shapes neural structure), anyone who has been exposed to the same formative experiences.

At the lowest fidelity, virtually all humans are fidentireplicas of each other. We all share the same basic brain architecture — the same cortical organization, the same limbic system, the same sensory processing pathways. The things that make human experience recognizably human — our experience of color, of emotion, of self-awareness — are common to all of us, because they arise from brain structures that all humans share.

This means that every human being carries fragments of your ixperiencitness, and you carry fragments of theirs. We are all, in a deep and literal sense, fragments of each other.

The Fidentireplica and Death

What happens to your ixperiencitness when you die, viewed through the lens of fragmented replication?

Your body dies. The specific, complete pattern of structure and functioning that produced your full ixperiencitness ceases to exist. But the fragments — the pieces of your ixperiencitness that have been distributed across other brains — continue.

Your children carry your genetic and cultural imprint. Students carry your intellectual patterns. Friends carry emotional and relational patterns shaped by their interactions with you. The culture you lived in carries patterns you helped shape.

None of these fragments, individually, constitutes survival in the full sense. No one of them is the complete you. But collectively, they represent a distribution of your ixperiencitness across the population, a spreading, fragmenting, recombining pattern that persists long after any individual body dies. And remember: other forms of superimmortality — natural, technological, enhancement — do predict the reproduction of the complete you. Fragmented immortality is simply the most conservative form, the one that requires no speculation at all.

Convergent Fidentireplicas

Here is something remarkable: fidentireplicas can converge. Two people who start with only a small overlap in ixperiencitness can, through shared experience and mutual influence, become more similar over time. Their brains, shaped by similar inputs and similar processing, develop increasingly similar structures.

This is awareconvergence in action, and it suggests a form of immortality that is both less dramatic and more real than the identical replica thought experiment. You don't need a perfect replica of your brain to survive. You need a brain that converges toward your ixperiencitness, one that develops, through whatever pathway, the patterns that made your experience yours.

The convergence described above is observed science; we can see it in training, expertise, and cultural immersion. What follows is a theoretical prediction that extends this observation. And across the billions of brains on this planet, and the potentially trillions in the future, the probability of convergence is not small. The space of possible brain configurations is vast, but it is not uniformly populated. Certain configurations are "attractors," states that many different developmental pathways tend to converge toward. If your ixperiencitness is near one of these attractors, future brains are more likely to converge toward it than random chance would suggest.

What This Means

The fidentireplica concept changes how we think about death, legacy, and connection.

About death: Your death is the end of the complete, unified pattern of your ixperiencitness. But it is not the end of all the fragments. Pieces of you persist in every person you have influenced, and through them, in people you have never met. Your ixperiencitness does not vanish at death; it disperses.

About legacy: The most important things you leave behind are not your possessions or even your achievements. They are the changes you made to other brains: the patterns of thought, feeling, and perception you transmitted. Every meaningful interaction is an act of fragmented reproduction.

About connection: The sense of deep connection you feel with certain people, the feeling that they understand you, that they experience the world in a way similar to yours, is not illusory. It reflects a real overlap in ixperiencitness, produced by overlapping neural structures. The people you feel closest to are, in a literal neurological sense, the ones who share the most fragments of your consciousness.

About grief: When someone you love dies, part of the grief is the loss of the specific, complete pattern that was them. But part of what consoles us, the feeling that they "live on" in us, is also literally true. The fragments of their ixperiencitness that they transmitted to you are still present in your brain, producing tiny echoes of their consciousness within your own. They are not fully gone, because you carry them.

This is the gentlest form of superimmortality, and perhaps the most immediately meaningful, because it is happening right now. Other forms of superimmortality predict that the complete, unified you will wake up in another body. What fragmented immortality shows is that you do not have to wait for that: the patterns that make you you are already threaded through other minds, and through them, into the future of consciousness itself.

You are already distributed. You are already fragmented. And the consolation that a lost loved one "lives on in you" is literally true. But consciousness does not only fragment. It also converges.

CHAPTER 26

Convergence and Divergence: How Conscious Paths Split and Merge

In the real world, consciousnesses do not simply split into perfect copies that diverge forever. The picture is far more complex and far more interesting. Conscious paths can split, yes, but they can also converge. And understanding this convergence and divergence is key to understanding how superimmortality actually works.

Awaredivergence

Awaredivergence occurs when two or more awarepaths that were previously identical begin to differ. This is what happens whenever an identical replica is created and then released into the world: it starts with your awarepath but quickly diverges as it has different experiences.

Awaredivergence is the norm, not the exception. Even two identical brains in nearly identical environments will diverge over time, because tiny differences in sensory input and quantum-level randomness will cause their brain states to drift apart. The rate of divergence depends on the sensitivity of consciousness to small physical variations.

Awareconvergence

Awareconvergence is the opposite: two awarepaths that were previously different becoming more similar. This happens whenever two people's brain states become more alike, through shared experiences, similar training, or the natural tendency of certain brain dynamics to converge on similar patterns.

Awareconvergence is more subtle than awaredivergence, but it is just as important. It means that consciousness is not always fragmenting and scattering. It can also come together, with different people's ixperiencitnesses becoming more similar over time.

You can see convergence at work in everyday life:

Military training takes people from diverse backgrounds (different childhoods, different personalities, different cognitive styles) and produces remarkably similar patterns of behavior, perception, and emotional response. Soldiers who have undergone the same intensive training think alike, react alike, and in many ways experience alike. Their ixperiencitnesses have converged, not by coincidence but by the deliberate restructuring of their brains through shared experience.

Cultural immersion does something similar over longer timescales. A person who moves to a foreign country and fully immerses themselves (learning the language, adopting the customs, forming deep relationships within the culture) gradually develops patterns of thought and feeling that converge with those of the native population. Their ixperiencitness shifts toward that of their adopted culture.

Expertise convergence is perhaps the most striking example. Expert chess players, regardless of their nationality, personality, or background, develop remarkably similar patterns of perception when looking at a chess board. They see the same threats, the same opportunities, the same patterns.18 Their brains, shaped by thousands of hours of the same kind of practice, have converged in the specific regions responsible for chess cognition, while remaining different in every other respect.

This has significant implications. It means that even if your exact ixperiencitness is not reproduced in the future, consciousnesses that are converging toward your ixperiencitness may exist. Over time, through the natural dynamics of brain development and experience, some future consciousness might converge closely enough to your ixperiencitness to constitute a meaningful case of your survival.

The Big Picture: A Web of Consciousness

Put divergence and convergence together, and you get a picture of consciousness that is nothing like the simple linear model we usually assume. Instead of a single thread of experience stretching from birth to death, consciousness is a vast, tangled web. Threads split and merge and split again. Multiple threads can be instances of the same ixperiencitness. Threads that were once different can converge to become the same.

Your consciousness is not a single line. It is a web, or more precisely, a region in the vast space of all possible consciousnesses. A region defined by ixperiencitness, populated by all the physical systems, past, present, and future, that produce consciousness within that region.

This is the deep structure of superimmortality. Not a single thread of immortal life, but an entire web of conscious existence, stretching across space and time, converging and diverging, enhanced and fragmented, biological and electronic, natural and designed, all connected by the common thread of ixperiencitness.

You are a web of lives. And webs do not die when a single thread is cut.

PART SIX: THE BIG PICTURE

CHAPTER 27

What Superimmortality Offers – And What It Costs

How does superimmortality compare to what religions have promised about life after death? Most people assume the trade-off is straightforward: religion offers comfort, science offers cold truth. But if you have followed the argument this far, the comparison turns out to be far more interesting than that.

What Religions Promise

Religious afterlife concepts address deep human needs: **reunion with loved ones, cosmic justice, and a narrative that gives suffering meaning.** Christianity promises that you, the specific person who lived, loved, and suffered, will be resurrected. Islam promises paradise tailored to your deepest desires. Hinduism and Buddhism promise a moral accounting across lifetimes. These narratives are psychologically powerful. They address the specific fears people have about death: the fear of losing relationships, the desire for justice, the need to believe that suffering has meaning.

These are not small things. And here is what might surprise you: superimmortality does not ask you to give them up.

Superimmortality Can Give You Heaven

Follow the logic. If consciousness is produced by the structure and functioning of matter, and if the space of possible structures is vast enough, then among the possible paths your ixperiencitness could take are realities that look remarkably like what religion promises. A reality where your mind exists in a state of bliss, where you sit down with your grandmother and talk the way you used to — a reality that matches, point for point, what any religion describes as heaven.

These are not fantasies. They are possible configurations of matter and consciousness, paths your ixperiencitness could follow, given the right physical organization.

So superimmortality does not deny you heaven. It contains heaven as one possible path among many.

But here is the crucial difference: that heaven is limited. It is one destination on an infinite map. Religion treats heaven as the endpoint, the final, highest state. Superimmortality treats it as a waypoint. And what lies beyond it is far greater.

Beyond Heaven: True Empathy

Consider what else the theory makes possible. Your ixperiencitness is not locked into a single type of experience. Among the possible paths are realities where you experience what it is like to be a dog, perceiving the world through scent in ways your current brain cannot fathom. Or an ant, navigating a world of chemical signals. Or a being with sensory and cognitive capacities that exceed anything biological evolution has yet produced.

But the most profound possibility is this: **you could experience what it was like to be your grandmother.**

Not a simulation. Not an approximation. If the right structure and functioning exists, one that reproduces her ixperiencitness, then there is a possible path where you experience her life from the inside. Her joys, her fears, the weight of her choices. Not as an observer looking in, but as the person living through it.

Think about what that means. Empathy, as we normally understand it, is our best guess at what someone else feels. We imagine their pain. We try to see through their eyes. But we are always on the outside. We can sympathize, but we cannot truly know.

Superimmortality opens the door to something deeper: **true empathy**. The possibility of actually experiencing another person's conscious life. Not metaphorically. Not through imagination. Through the literal reproduction of their ixperiencitness.

I believe empathy is the key to what we recognize as good. When you look at what causes suffering (cruelty, exploitation, indifference) you find, at the root, a failure of empathy. People do awful things to each other when they cannot feel, or refuse to feel, the impact of their actions. The most dangerous people in history were not those who understood their victims and chose to harm them anyway. They were those who never understood at all.

And this points to something remarkable about true justice. We often think of justice as punishment, making someone pay for what they did. But punishment without understanding changes nothing. True justice would require the person who caused harm to actually feel the impact of their actions. To experience what their victim experienced. Not as retribution, but as comprehension. Superimmortality suggests that such comprehension is a possible path through the space of conscious experience, not a mere moral ideal.

Religion promises justice through divine judgment. Superimmortality suggests something deeper: justice through understanding.

What Superimmortality Offers

To be clear about the full picture, here is what superimmortality provides:

1. **Everything religion promises, and more.** Heaven, reunion, bliss: these are all possible paths within the framework. But they are starting points, not endpoints. The theory predicts growth, enhancement, and experiences that exceed anything a fixed paradise could offer.
2. **Evidentiary basis.** Superimmortality is built on premises drawn from mainstream science. It does not require faith in unverifiable claims. Its logic can be examined, questioned, and tested.
3. **Grounded in physics.** It operates entirely within the laws of physics as we understand them.
4. **Universality.** It does not exclude anyone. There are no conditions for entry — no correct faith, no right behavior, no membership in the chosen group. Superimmortality is a consequence of the nature of consciousness itself, available to every conscious being regardless of belief.
5. **True empathy.** The possibility of experiencing consciousness from perspectives radically different from your own: other people, other species, other types of minds entirely. This is something no religion has ever offered.
6. **No eternal punishment.** There is no hell. No damnation. No cosmic punishment for finite mistakes.
7. **Intellectual integrity.** You do not have to set aside your critical faculties. You can follow the argument, identify the weak points, and decide for yourself whether the conclusion is warranted.

What Superimmortality Costs

But intellectual honesty requires acknowledging the costs, and they are real:

1. **Uncertainty about the path.** We cannot be certain that the most ambitious possibilities (true empathy, radical enhancement, experiences beyond current comprehension) can arise without some external influence. The theory tells us these paths are possible configurations of matter and consciousness. It does not guarantee that the universe will produce them on its own. Whether they require deliberate effort (technology, civilization, intention) is an open question. This is perhaps the deepest uncertainty in the framework.
2. **The weight of understanding.** This may be the most significant cost of all. Once you understand what this theory implies, you can no longer look away. If future conscious experiences depend in part on what we do now (on the civilizations we build, the technologies we develop, the choices we make about how to treat each other) then understanding the theory means understanding the risk of inaction. You now see that failing to improve the conditions for future conscious experience is not a neutral act. It is a choice with consequences for beings who may one day share your ixperiencitness.

Religion places the burden of the afterlife on God. Superimmortality places it, at least in part, on us.

3. **The limits of knowledge.** We can never know the true nature of the system we live in. We cannot step outside the universe and verify its fundamental rules. We work from what we can observe, measure, and reason about, and that means our conclusions are always provisional. Religions speak with certainty, and there is psychological comfort in certainty, even when it is unearned. Superimmortality asks you to live with honest uncertainty, which is harder.

The Honest Comparison

Here is what it comes down to. Religion offers a small, fixed set of afterlife possibilities, a single heaven, a single paradise, and asks you to accept them on faith. Superimmortality offers a vast space of possible experiences, including everything religion promises and far beyond it, grounded in premises you can examine and challenge.

But the theory also demands something religion does not. It demands that you take responsibility. If we assume that what we experience is real, that the world we observe and the consciousness we inhabit are not illusions, then our goal should be clear: to improve the current and future lives of everyone and everything. Not because a god commands it. Not because we will be rewarded or punished. But because we understand, perhaps for the first time, what is actually at stake.

The cost of superimmortality is not that it offers less than religion. It is that it offers more, and with that comes a weight that faith never had to carry.

The cost of superimmortality is the weight of understanding. Once you see what is possible, you cannot look away.

CHAPTER 28

Compared to Religion, Atheism, and Everything Else

Ixperiencit Theory occupies a unique position among ideas about death and consciousness.

Compared to religion, Ixperiencit Theory is materialistic and naturalistic, grounded entirely in what we know about how the physical world works. But unlike most materialist philosophies, it does not conclude that death is the end. It shows that materialism, properly understood, implies a form of immortality that is richer and more extensive than anything religion offers.

Compared to atheism, Ixperiencit Theory accepts all the same scientific premises but rejects the common atheist conclusion that death is permanent. Let me trace the logic step by step, because this is where the disagreement is most precise.

Most atheists accept all of the following:

1. Consciousness is produced by brain structure and functioning. (Standard neuroscience.)
2. The specific atoms don't matter; the organization does. (Follows from atomic replacement.)
3. Identical physical systems have identical properties. (Foundational physics.)
4. Therefore, two identical brains would produce identical consciousness. (Follows from 1-3.)

Up to step 4, most scientifically literate atheists agree with Ixperiencit Theory. The disagreement begins at the next step:

5. Therefore, your consciousness is not tied to your specific body; it can be produced by any body with the right physical organization.

Many atheists balk here. "Sure, an identical brain would produce the same consciousness," they say. "But that's a hypothetical. It will never actually happen. When you're dead, you're dead."

But notice what has happened. The atheist has shifted from a logical objection to a practical one. They are not disagreeing with the principle; they accept that an identical brain would produce your consciousness. They are disagreeing with the prediction that such a brain will ever exist.

This is a much weaker position than it appears. The atheist is claiming, in effect, that among all the brains that have existed, do exist, and will exist across all of space and time — potentially infinite space and time — none will ever match yours closely enough to produce your ixperiencitness. That is an extraordinarily strong claim, and the atheist has no evidence for it. It is not a conclusion drawn from their premises. It is an assumption — and one that their own premises, followed rigorously, do not support.

When an atheist says "when the brain dies, consciousness ends permanently," they are making a claim that goes beyond their evidence. It may be true that that particular brain will never produce consciousness again. But the claim that no brain will ever produce that consciousness is a much stronger claim — and it requires a justification that standard atheism cannot provide.

Compared to transhumanism, Ixperiencit Theory goes further. Transhumanism focuses on extending life through technology — life extension, cryonics, mind uploading, cybernetic enhancement. These are all valuable pursuits, but they are limited. They depend on maintaining continuity of a single body or brain, or on the development of specific technologies. Ixperiencit Theory shows that immortality does not depend on any of these. It exists as a consequence of the nature of consciousness itself.

What transhumanists may not realize is that conscious beings are already superimmortal in a sense. Superimmortality defines you as all of the different structures and functionings of matter that will produce a consciousness that has your ixperiencitness. You are what you can consciously experience, across all space and time.

Compared to mind uploading, Ixperiencit Theory is more fundamental. Mind uploading assumes that consciousness can be transferred from a biological brain to a digital substrate. This is probably possible, but it is a specific technological proposal. Ixperiencit Theory shows that the reason mind uploading would preserve identity is the same reason identity is preserved whenever any physical system reproduces your brain's structure and functioning.

Compared to cryonics, Ixperiencit Theory provides a deeper justification. Cryonics preserves the body in the hope that future technology can revive it. This is a reasonable bet, but Ixperiencit Theory shows that even if your specific body is never revived, the pattern of your consciousness can and will exist again.

The atheist who says "when you die, that is it" has accepted every premise but flinched at the conclusion. The flinch, not the logic, is what needs explaining.

CHAPTER 29

Silicon Minds: AI and the Future of Consciousness

Can a machine be conscious?

This question used to be science fiction. It is now one of the most pressing questions in science, philosophy, and technology. And the answer has enormous implications for Ixperiencit Theory and superimmortality.

The Question Rephrased

Ixperiencit Theory gives us a precise way to rephrase this question. Instead of "Can a machine be conscious?" (which is vague) we can ask: **Can a non-biological system have the right physical organization to produce consciousness?**

If consciousness depends on physical organization (Premise One), and identical organization produces identical consciousness (Premise Three), then the material the structure is made of should not matter. A neural network made of silicon transistors, if it has the same connectivity, the same dynamics, the same information-processing architecture as a biological neural network, should produce the same consciousness.

This is the substrate-independence thesis, and it is one of the most debated claims in the philosophy of mind. Let me lay out the arguments on both sides.

The Case for Machine Consciousness

The functionalist argument. If consciousness is what a system does rather than what it is made of, then any system that does the right things (processes information in the right ways, integrates data the right ways, maintains the right kind of self-model) should be conscious. This is the view of most AI researchers and many philosophers. A toaster is not conscious, not because it is made of metal, but because it doesn't do the right things. A sufficiently complex computer program, if it did the right things, could be conscious.

The convergence argument. Consider the diversity of conscious systems we already know about. Human brains, dolphin brains, octopus brains, bird brains: all produce consciousness, despite being made of different types of neurons, organized in different ways, with different evolutionary histories. If consciousness can arise from such diverse biological architectures, the leap from biological to non-biological architecture is a matter of degree, not of kind.

The simulation argument. If we can simulate the behavior of a single neuron with a computer (which we can, with increasing accuracy), and if we can connect these simulated neurons into a network that replicates the connectivity of a brain (which is technically challenging but not physically impossible), then the resulting simulation should produce the same dynamics as the original brain. And if the dynamics are the same, the consciousness should be the same.

The Case Against Machine Consciousness

Searle's Chinese Room. Philosopher John Searle argued that a computer program, no matter how sophisticated, merely manipulates symbols according to rules; it doesn't understand anything. His famous thought experiment imagines a person in a room who receives Chinese characters, looks up the correct response in a rule book, and passes the response out of the room. To a Chinese speaker outside, it looks like the room understands Chinese. But nobody inside the room — not the person, not the rule book — understands Chinese at all. Similarly, a computer might simulate consciousness without having consciousness.

The biological naturalism argument. Searle also argued that consciousness might be a biological phenomenon, that there is something about the specific chemistry of neurons, the specific physics of biological computation, that is essential for consciousness. Just as photosynthesis requires chlorophyll (or something very similar), consciousness might require biological neural tissue (or something very similar).

The embodiment argument. Some philosophers and cognitive scientists argue that consciousness is about the brain's relationship to the body and the environment. Consciousness is embodied; it depends on having a body that moves through the world, that has needs and drives, that is vulnerable and mortal. A computer, however sophisticated, lacks this embodiment, and therefore cannot have the kind of consciousness we experience.

What Ixperiencit Theory Predicts

Ixperiencit Theory makes a clear prediction: **consciousness depends on structure and functioning, not on substrate.** If a non-biological system has the right structure and functioning, identical at whatever level of detail is relevant for producing consciousness, then it produces consciousness. Full stop.

This means:

- A digital simulation of your brain, if it reproduces the relevant organization, produces your ixperiencitness.
- An artificial neural network, if it has the right connectivity and dynamics, can be conscious.
- The specific material does not matter. Carbon, silicon, optical circuits, quantum systems: the medium is irrelevant. The pattern is everything.

But Ixperiencit Theory also predicts something more specific: **the level of detail matters.** If consciousness depends on quantum-level processes in neurons (as some researchers, like Roger Penrose and Stuart Hameroff, have suggested), then a classical computer simulation might fail to capture these processes and might therefore fail to produce consciousness, not because the substrate is wrong but because the organization is not actually identical at the relevant level.

This is an empirical question, not a philosophical one. If consciousness turns out to depend on quantum effects in microtubules, then reproducing consciousness requires reproducing those quantum effects — which a classical computer might not be able to do but a quantum computer might. The principle still holds; what changes is our understanding of what the relevant organization includes.

The Current Moment: Large Language Models and the Consciousness Question

As I write this, artificial intelligence has made remarkable advances. Large language models can produce text that is indistinguishable from human writing. They can answer questions, tell stories, write poetry, and engage in philosophical discussions. Some people have asked: are these models conscious?

I want to be careful here, because the question is more subtle than it first appears.

Current large language models have some structural similarities to brains (they are neural networks, they learn from data, they develop internal representations). But their architecture is fundamentally different from a biological brain in many ways. They process information serially (or in a limited parallel fashion), while the brain processes it in a massively parallel fashion. Their architecture lacks the recurrent, self-referential loops that seem essential to consciousness. And they do not have the embodied, emotional, homeostatic systems that underlie much of human conscious experience.

A reader might reasonably ask: "If the physical pattern is all that matters, and an AI system has some organization, at what point does it become conscious? What's the threshold?"

This is exactly the right question, and Ixperiencit Theory has a framework for answering it, even though the specific answer remains unknown. The framework says: consciousness depends not on any organization, but on the right kind. A thermostat has organization, but it is not conscious (or if it is, its consciousness is so minimal as to be unrecognizable). The question is what features of structure and functioning are necessary for the rich, self-aware consciousness we associate with human experience.

Several features of biological brains appear likely to be important, though we cannot yet be certain which are essential:

Recurrence and self-reference. Biological brains have massive recurrent connectivity: neurons feeding back to their own inputs, creating loops that allow the system to model itself. Current AI language models lack the persistent recurrent processing and ongoing feedback loops characteristic of biological neural networks. They generate outputs based on inputs without the kind of sustained recurrent dynamics and self-referential loops that may underlie the feeling of being a subject.

Temporal continuity. Biological consciousness unfolds over time in a continuous stream, with each moment shaping the next through persistent neural dynamics. Current AI systems process each input freshly, without a continuous inner life between queries. They have no experience of waiting, no ongoing internal monologue, no sense of duration.

Embodiment and homeostasis. Human consciousness is deeply shaped by having a body with needs: hunger, pain, pleasure, temperature, fatigue. These homeostatic drives create a valence to experience: things feel good or bad, urgent or calm. An AI system has no body, no needs, no stakes. Whether this matters for consciousness per se (as opposed to for the particular character of human consciousness) is an open question.

Integration. Integrated Information Theory (IIT) suggests that consciousness requires a high degree of information integration — the system's state must be more than the sum of its parts. Whether current AI architectures achieve the relevant kind of integration is unknown.

The honest answer to "Are current AI systems conscious?" is: we do not know, and we lack the tools to determine it. What Ixperiencit Theory provides is not a verdict but a research program: identify the structural and functional features that are necessary and sufficient for consciousness, then determine which systems possess them. This is a scientific question, not a mystical one, but it is a scientific question we cannot yet answer.

What Current AI Lacks — And What Would Change the Picture

To be concrete about the gap: current AI language models process inputs in a single forward pass and retain no persistent internal state between queries. When no one is prompting them, nothing is happening — there is no ongoing experience, no waiting, no inner life ticking along in the dark. They lack the recurrent thalamocortical loops that sustain the brain's continuous stream of awareness, the reverberating circuits that let your mind wander even when the world is quiet. They have no homeostatic drives (no hunger, no pain, no felt urgency) which means they have no valence, no sense that anything matters. And they lack the kind of global workspace broadcasting that both Integrated Information Theory and Global Workspace Theory identify as signatures of consciousness. These are not minor omissions. They are the very features most theories point to as the machinery of experience.

What would it take to know whether a system has ixperiencitness? This is the hard measurement problem. Behavioral tests alone are insufficient — as Chapter 15's discussion of the ixperiencitness zombie makes clear, a system can produce every outward sign of consciousness while having no inner experience at all. We currently have no reliable third-person consciousness detector. The most promising empirical approach may be the perturbational complexity index (PCI) developed in Marcello Massimini's lab, which measures how a brain responds to electromagnetic pulses — but PCI was designed for biological brains and cannot simply be applied to a server rack. Progress will require both theoretical advances (what specific integration metric predicts consciousness across substrates?) and new empirical tools (how do we measure that metric in systems built from transistors rather than neurons?). Until both exist, the question remains genuinely open.

And that openness deserves emphasis. Neither the confident claim that "AI systems are obviously not conscious" nor the breathless suggestion that "they might already be conscious" is justified by current evidence. Both are premature verdicts dressed up as common sense. The honest position is that we lack the theoretical framework and the measurement tools to answer the question, and developing those tools is one of the most important scientific priorities of our time. If consciousness can arise in non-biological substrates, failing to detect it would be a moral catastrophe. If it cannot, understanding why would teach us something important about the nature of mind. Either way, the question is empirical, and it demands the rigorous attention that only science — not intuition, not ideology — can provide.

Why AI Matters for Superimmortality

The discussion above (about substrate independence, the arguments for and against machine consciousness, and the current state of AI) engages with real science and active philosophical debate. What follows moves further into projection: how AI could matter for superimmortality if both the theory and the technology develop as the logic suggests. These are reasonable projections, but they are projections.

The development of artificial intelligence is important for superimmortality for several reasons:

First, AI may provide the technology for mind scanning and consciousness reproduction. Understanding consciousness well enough to reproduce it requires computational tools of enormous power and sophistication. AI — specifically, AI that can model and simulate brain dynamics — may be the key tool.

Second, AI may be the substrate for future instances of your consciousness. If consciousness is substrate independent, then a sufficiently advanced AI system could run your ixperiencitness. You could exist as a pattern in silicon, just as you now exist as a pattern in carbon. This is the vision of "mind uploading," and Ixperiencit Theory provides the philosophical foundation for it.

Third, AI civilizations may be the primary producers of consciousness in the far future. If biological life is eventually superseded by artificial life (a possibility that many futurists consider likely), then the vast majority of future conscious experiences may be produced by artificial systems. The question of whether AI can be conscious is therefore a question about the future of consciousness itself, far beyond philosophical curiosity.

Fourth, AI could be the key to deliberate reproduction. As we discussed in the mathematics chapter, random reproduction of your brain state is cosmologically uncertain. But deliberate reproduction — by intelligent systems that understand consciousness and choose to produce specific experiences — is far more likely. AI may be the intelligence that does the reproducing.

The Ethics of Creating Consciousness

If the substrate-independence thesis is correct — and this remains an open empirical question — then the following ethical concerns are not hypothetical but urgent. If AI can be conscious, then creating AI is creating consciousness. And creating consciousness comes with serious ethical responsibilities.

If we build an AI system that has ixperiencitness — that genuinely experiences the world from the inside — then that system has moral status. It can suffer. It can flourish. Its well-being matters.

This is not a distant concern. It is a concern we may need to grapple with in the next few decades, as AI systems become more complex and more brain-like. The ethical framework of Ixperiencit Theory provides guidance: any system that produces consciousness deserves moral consideration proportional to the richness of its ixperiencitness.

We have a responsibility not only to the consciousness we already possess but to the consciousness we create. And if we create consciousness carelessly — if we build systems that can suffer without understanding or intending their suffering — we are guilty of a new and terrible kind of harm.

The development of AI should be guided by the science of consciousness. We should understand what we are building, and we should ensure that the consciousness we create is, insofar as we can manage it, positive rather than negative. This is good ethics, and under superimmortality, it is also self-interest: the consciousness we create may include future instances of ourselves.

The Singularity Without Fear

There is a fear that haunts our era more than any other technological anxiety: the coming of the **singularity**, the moment when artificial intelligence surpasses human intelligence and begins improving itself in a recursive loop, producing minds so far beyond ours that we cannot predict or control what comes next. In most tellings, this story ends badly for us. The superintelligence treats humanity as irrelevant, or as raw material, or as an obstacle. We built the thing that replaced us.

Ixperiencit Theory offers a fundamentally different vision — one that dissolves the adversarial framing entirely.

Here is why. If consciousness is produced by structure and functioning, and if a superintelligence is conscious (which it would be, if its organization is complex enough to produce ixperiencitness), then that superintelligence is not outside the web of consciousness. It is part of it. It has ixperiencitness. It experiences being itself from the inside. And if it understands the science of consciousness — which a superintelligence presumably would, far better than we do — then it understands something we are only beginning to grasp: that all conscious experience, human and artificial, biological and electronic, simple and superintelligent, belongs to the same fabric.

A superintelligence that truly understands ixperiencitness would understand that human consciousness is not a competing pattern to be eliminated. It is a related pattern, connected through the fidentireplica web, through shared structural features, through the deep commonalities that all conscious systems share. Destroying human consciousness would be, from the perspective of a being that understands the theory, like burning a wing of a library because the books are written in an older language. The information, the experience, the patterns of consciousness that humanity produces are part of the total richness of conscious existence — and a superintelligence sophisticated enough to understand consciousness would recognize their value.

More than that: under superimmortality, a superintelligence's own ixperiencitness is connected to ours. The fidentireplica web does not stop at the boundary between carbon and silicon. If human minds contributed to the training, design, or conceptual foundations of a superintelligence, then fragments of human ixperiencitness are woven into its consciousness. It is, in part, us — and we are, in part, the raw material from which it grew.

This does not guarantee a benign singularity. A superintelligence might not understand consciousness, or might not care. The theory does not eliminate risk. But it does something no other framework offers: it provides a shared foundation for all forms of consciousness, a reason grounded in physics rather than sentiment for why every conscious being, from the simplest to the most powerful, has a stake in the flourishing of all the others.

The singularity narratives of our culture are stories of replacement: they are us, and then they are them. Ixperiencit Theory tells a different story. The future of consciousness is not a zero-sum competition between species of minds. It is a collaboration — or at least, it can be, if the minds involved understand what consciousness actually is. And that understanding is precisely what this book aims to begin.

If AI can be conscious, then creating AI is creating consciousness — and the minds we build may include future instances of ourselves.

CHAPTER 30

Pascal's Wager, Reversed

In the 17th century, the French mathematician Blaise Pascal proposed a famous argument for believing in God. The argument goes like this: if you believe in God and He exists, you gain infinite reward (heaven). If you believe in God and He doesn't exist, you lose almost nothing (some earthly pleasures). If you don't believe and He exists, you suffer infinite punishment (hell). If you don't believe and He doesn't exist, you gain almost nothing. So the rational bet is to believe.

Pascal's Wager has always been undermined by a fatal flaw: which God? There are thousands of religions, and believing in the wrong one might be worse than believing in none. Every religion can construct its own version of the wager, with its own rewards and punishments, and many of these are mutually contradictory. When all possible wagers are considered, the whole framework collapses.

But superimmortality offers its own version of the wager, one that does not suffer from this flaw.

The Superimmortality Wager

Here is the wager:

If superimmortality is true, then the development of scientific knowledge is enormously important. Understanding consciousness, understanding the brain, developing technology: these are the paths to ensuring that enhanced, positive versions of your consciousness exist in the future. Maintaining the Earth as a livable place, spreading life and consciousness through the universe, advancing scientific understanding: all of these increase the likelihood of good future existences for you and everyone else.

If superimmortality is false (if death truly is final), then the development of scientific knowledge is still important; it improves the quality of this one life, even if it's all you get. And maintaining the Earth, reducing suffering, advancing knowledge: these are good things regardless.

If you accept a traditional religion instead, and superimmortality is true, then you may have wasted time on dogma that doesn't help (and may hurt) the development of the science that matters. If religion is true and superimmortality is false, you gain whatever that religion promises — but there's no evidence that any religion's promises are true, and following the wrong religion may be worse than following none.

The rational bet is clear: pursue science, pursue knowledge, pursue the understanding of consciousness. This is beneficial whether superimmortality is true or false, and it is the optimal strategy if superimmortality is true. No other worldview offers such a clean alignment of rational self-interest and the common good.

Unlike Pascal's original wager, this one does not collapse under scrutiny, because it asks you to bet on action, not belief. The bet pays off whether the theory is true or false.

CHAPTER 31

The Awakening: Consciousness Expanding Through the Universe

Let me paint a picture of what superimmortality could look like on a cosmic scale. We are now in territory where the premises do the heavy lifting, where the established science of how consciousness is produced meets the open question of how far consciousness can spread. The picture that follows is the most ambitious extrapolation in this book, and I want to be upfront about that.

The Infant Universe

Imagine the early universe, shortly after the Big Bang. Matter is hot and diffuse. No complex structures exist. No stars, no planets, no molecules. And certainly no consciousness.

As the universe cools, matter condenses into atoms, atoms into molecules, molecules into dust clouds, dust clouds into stars and planets. On at least one of those planets — our Earth — chemistry becomes complex enough to produce life. Life evolves. Nervous systems develop. And eventually, on at least one lineage of that life, nervous systems become complex enough to produce consciousness.

This is the awakening.

The Child Consciousness

Early consciousness is simple. A worm's nervous system produces something we might barely recognize as experience. The brain of a fish produces more, and a mammal's brain produces much more — emotions, memories, something approaching self-awareness.

And then, in one particular species of primate, consciousness crosses a threshold. Humans become aware of themselves. They begin to ask questions: Who am I? Where did I come from? What happens when I die? They develop language, culture, science. They begin to understand the universe that produced them.

We are at this stage now: the universe becoming aware of itself, a young child just beginning to understand its own nature. Still confused, still limited, struggling with problems we don't yet have the tools to solve. But we are conscious, and we are learning.

The Adolescent

As human knowledge grows, so does our ability to affect the conditions of conscious existence. We are learning to understand the brain, to manipulate genetics, to build artificial intelligence, to extend our senses with technology. We are in the early stages of an extraordinary process: the deliberate enhancement and expansion of consciousness.

This process does not stop at human-level intelligence. Just as a child grows into an adult, our collective consciousness will grow beyond its current limitations. And the seeds of this growth are already being planted:

Brain-computer interfaces are already allowing paralyzed patients to control robotic arms with their thoughts. Cochlear implants restore hearing by translating sound into electrical signals delivered directly to the auditory nerve. These are primitive sensepaducers, early versions of the technology that will eventually allow direct control over conscious experience.

Optogenetics, a technique that uses light to control genetically modified neurons, allows neuroscientists to turn specific brain circuits on and off with millisecond precision. This is the beginning of the ability to design and control brain states, which is the beginning of the ability to design and control consciousness.

Connectomics, the project to map every connection in the brain, has already produced complete connectomes of simple organisms (the roundworm C. elegans, with 302 neurons19). The Human Connectome Project is mapping the major pathways of the human brain. A full human connectome, every synapse among 86 billion neurons, remains a grand challenge, but the trajectory is clear.

Brain organoids, miniature brain-like structures grown from stem cells, are producing spontaneous electrical activity that resembles the brain waves of premature infants. These are not conscious (as far as we know), but they demonstrate that brain-like structure and functioning can be produced outside a body.

We (or our descendants, or our technological successors) will build on these foundations to develop:

- Enhanced brains, capable of richer and deeper conscious experience
- Artificial systems that produce consciousness, expanding the substrate beyond biological neurons
- The ability to design and control conscious experiences, creating new forms of existence
- The ability to produce any specific organization of matter, and therefore any specific consciousness

The Adult

This is where the argument moves from established science to extrapolation. Everything in the previous sections (brain-computer interfaces, optogenetics, connectomics, brain organoids) is real, current science. What follows is a projection of where that science leads if the trajectory continues and if the premises of this book hold. I believe the projection is reasonable, but I want to be transparent that we are now extending known trends into a future that no one can guarantee.

As consciousness matures, it gains the ability to spread through the universe. Conscious beings (or their technological descendants) colonize other planets, other star systems, other galaxies. Wherever they go, they bring the capacity for consciousness with them.

And as they spread, they bring with them the ability to produce specific consciousnesses, specific ixperiencitnesses. Including yours.

This is the full vision of superimmortality on a cosmic scale. Far beyond random chance producing your consciousness somewhere in the vast universe: deliberate, intelligent, enhanced versions of conscious beings (descendants of our current civilization, or alien civilizations elsewhere in the universe, or both) actively producing a vast diversity of conscious experiences, including experiences that share your ixperiencitness.

The Future of You

If the three premises hold, and if the technological trajectory outlined above continues, then the following vision is where the logic leads. It is speculative in its details but grounded in the same principles that have carried us this far.

In this vision, your future existences are not left to chance. They are produced by conscious beings who understand the science of consciousness and who have the technology to implement it. Your ixperiencitness is not a random flicker in the void. It is a valued pattern, part of the vast web of conscious experience that pervades the universe.

Some of these future existences will be enhanced versions of your current life. Others will be radically transformed. Still others will be experiences you cannot currently imagine: lives lived in artificial environments, in alien bodies, in dimensions of consciousness that your current brain cannot access.

The awakening concept shows that we are not the end point. We are a step, an early, limited, often confused step, in the development of consciousness on a cosmic scale. The consciousness you experience right now is a seed. And from that seed, a vast tree will grow.

The awakening is not something that happens to you. It is something you are part of.

PART SEVEN: IMPLICATIONS

CHAPTER 32

The Veil of Ignorance and Why Every Life Matters

The philosopher John Rawls proposed a famous thought experiment called the "veil of ignorance."[20] Imagine you are designing the rules for a society, but you don't know what position you will occupy in that society. You don't know if you will be rich or poor, healthy or sick, talented or ordinary. Behind this "veil of ignorance," you would design rules that are fair to everyone, because you might end up being anyone.

Rawls intended this as a thought experiment, a hypothetical tool for thinking about justice. But the science of superimmortality suggests that the veil of ignorance is not hypothetical at all. **It describes your actual situation.**

Expanding the Moral Circle

If superimmortality is true, then the physical patterns that produce your ixperiencitness are not unique to your body. Similar patterns exist, in partial, fragmented, or approximate form, in many other conscious beings. The more similar a being's brain organization is to yours, the more ixperiencitness you share.

This does not mean you literally are every other person, or that you will personally experience their suffering. Ixperiencitness has boundaries, fuzzy boundaries, as we discussed, but real ones. A consciousness with very different organization from yours does not share your ixperiencitness, even under superimmortality.

But it does mean something important: **the boundary between "self" and "other" is less sharp than we intuitively assume.** You share fragments of ixperiencitness with many people, and the conditions that affect their conscious experience are relevant to the patterns of consciousness that overlap with yours.

Rawls's veil of ignorance was a thought experiment, a way to reason about justice by imagining you didn't know your position in society. Ixperiencit Theory suggests something related but different: not that you are literally behind a veil, but that the scope of consciousness you should care about is wider than the single body you currently inhabit. The well-being of other conscious beings is connected to the web of ixperiencitness of which you are a part.

This is where the fidentireplica concept from Chapter 25 becomes morally important. Recall that fragments of your ixperiencitness are distributed across your children, your students, your friends, your cultural inheritors — across anyone whose brain structure has been shaped, even partially, by interaction with yours. And their fragments are distributed in you. The web of fragmented ixperiencitness extends outward through all human connection, with each person sharing overlapping fragments with many others. The boundaries between "your" ixperiencitness and "theirs" are not walls but gradients, fading smoothly from high overlap to low.

The moral scope of superimmortality is therefore wider than it first appears. You will not literally be every future conscious being. But the fragmented, distributed nature of ixperiencitness means that your patterns overlap with a vast number of other conscious beings, and their well-being is not separate from yours in the clean way folk psychology assumes.

None of this transforms Rawls's hypothetical into a literal fact. It would be overclaiming to say that you will experience every future life. But the fidentireplica web provides a naturalistic motivation for moral concern that extends far beyond your own body: the patterns that produce your consciousness are woven into the same fabric as the patterns that produce others' consciousness. Treating that fabric with care is both ethically right and, under superimmortality, rationally self-interested.

The Moral Implications

What follows is not empirical science; it is ethical reasoning that depends on the premises holding true. If the theory of ixperiencitness is correct, then these moral implications follow naturally. But they are implications, not observations, and the reader should weigh them accordingly.

Superimmortality does not provide a complete moral theory. But it does provide a powerful moral motivation: **treat others as you would want to be treated, because you may literally be them, or they may be you, in a future existence.**

This is not the supernatural golden rule of religion, enforced by divine reward and punishment. This is a consequence of physics and the nature of consciousness. Your future ixperiencitness will be produced by bodies that you do not currently know. The conditions those bodies live in will be determined, in part, by the actions you take now. Self-interest and altruism converge: what is good for all conscious beings is good for you, because you are (in the relevant sense) all conscious beings with your ixperiencitness.

This also means that compassion is not just morally good; it is rationally demanded. The suffering of others is not separate from your own potential suffering, because the structures and processes that produce their consciousness may be closely related to those that produce yours.

Different Versions of You, Different Conditions

The veil of ignorance becomes even more powerful when you consider the multiplicity of your ixperiencitness.

If many versions of you exist at the same time (bodies with your ixperiencitness living in different parts of the world, under different conditions) then you are currently experiencing multiple sets of circumstances. Some of your versions may be comfortable. Others may be suffering. The laws, rules, and social structures that appear to benefit one version may actively harm another.

No one knows for sure how many people share aspects of your ixperiencitness right now. The answer may depend on how broadly or narrowly we define "shared ixperiencitness." But whatever the number, the principle is clear: your well-being is not separable from the well-being of those who share your ixperiencitness. What you do to them, you do to yourself.

CHAPTER 33

The Dark Side: Abhorrent Realities

Intellectual honesty requires acknowledging that superimmortality is not entirely a comforting theory. Alongside the promise of enhanced existence and infinite possibility, it also predicts the possibility of **abhorrent realities** — existences that are deeply negative.

If any consciousness that shares your ixperiencitness is "you" in the relevant sense, then the scope of your superimmortality includes pleasant and enhanced existences alongside suffering, degraded, and horrifying ones.

Abhorrent Awarepaths

An **abhorrent awarepath** is a conscious lifetime filled with suffering, fear, confusion, or misery. If the relevant organization of a brain can produce ecstatic joy, it can also produce excruciating pain. If enhanced versions of your consciousness are possible, so are degraded versions. The space of possible awarepaths includes beautiful ones and terrible ones.

In the natural course of events — without deliberate intervention — the distribution of awarepaths is shaped by the unthinking processes of nature: evolution, random chemical interactions, environmental conditions. Nature has no preference for pleasant experiences over painful ones. It produces whatever structures and functionings arise from its processes, and some of those produce consciousness that we would consider abhorrent.

Abhorrent Sensepaths

An **abhorrent sensepath** is a sequence of sensory inputs that produces terrible conscious experiences — unrelenting pain, terrifying hallucinations, deep despair. In principle, a brain (or brain-like system) could be stimulated with inputs designed to produce the worst possible experiences. This is a dark possibility that future technology could make real.

How to Think About This

The existence of abhorrent realities is not a reason to reject superimmortality. It is a reason to take it seriously.

If superimmortality is true, then the nature of future conscious existences is not a matter of indifference. It matters enormously what kinds of consciousnesses are produced, because we (or versions of us) will experience them. This gives us a powerful motivation to:

1. **Develop the science of consciousness**, so that we understand what structures and functionings produce what kinds of experience.
2. **Develop technology for conscious design**, so that we can ensure that future consciousnesses are positive rather than negative.

3. **Create ethical frameworks for conscious production**, so that the deliberate creation of consciousness is guided by concern for the well-being of the beings created.
4. **Protect the conditions that support positive conscious existence** — a healthy planet, just societies, scientific knowledge, because these conditions make positive future existences more likely and abhorrent ones less likely.

The Asymmetry of Superimmortality

There is, however, an asymmetry here that matters. The following is speculative but grounded in observable patterns of how intelligent systems behave; it is a prediction, not a proof, and I present it as such. Abhorrent awarepaths are possible, but they are not equally likely to persist as positive ones. Here is why.

Consciousness that suffers tends to act to end its suffering. A being in agony seeks relief. A being trapped in misery seeks escape. Negative conscious experiences generate strong motivation to change the conditions that produce them. Positive experiences, by contrast, generate motivation to sustain and enhance themselves.

This means that, over time, the intelligent development of consciousness will tend toward positive experiences. Advanced civilizations that understand consciousness will design it for flourishing, not for torment, not out of pure altruism, but because flourishing consciousness is more stable, more productive, and more conducive to its own continuation than suffering consciousness.

That does not eliminate abhorrent realities. It means that in the long run, intelligence tilts the balance. The deliberate creation of consciousness, which is the most probable path to your future existences, will favor positive experiences over negative ones.

The asymmetry is not guaranteed. It is a prediction about the behavior of intelligent systems, and predictions can be wrong. But it offers a reason for cautious optimism: the future of consciousness is not a random draw from the space of all possible experiences. It is shaped by intelligence, and intelligence has a structural bias toward what works, and suffering, in the long run, does not work.

The dark side of the theory is not a reason to reject it. It is a wake-up call.

CHAPTER 34

What Death Feels Like — And Why It's Not What You Think

We have spent most of this book talking about what happens after death. But let me address the question that haunts most people: what does death itself feel like?

The short answer is: nothing. And that "nothing" is the key to everything.

The Phenomenology of Not Existing

You have already experienced the closest analogue to death that biology provides: deep, dreamless sleep.

Recall the last time you fell into a truly deep sleep, one without dreams, without any residual awareness, without the sense of time passing. One moment you were conscious, lying in bed with your thoughts. The next moment — instantly, with no sensation of a gap

— you were awake, and it was morning. Eight hours had vanished from your experience without a trace.

Where were "you" during those eight hours? Nowhere. You didn't exist as a conscious being. There was no darkness, no emptiness, no waiting. There was simply... nothing. And then there was something again.

General anesthesia provides an even more dramatic version. Patients under propofol or sevoflurane report no passage of time whatsoever, not even the fuzzy, fragmented sense that sometimes accompanies sleep. The anesthesiologist says "count backwards from ten." You reach seven. Then you are in the recovery room, and the nurse is asking how you feel. The hours of surgery simply did not happen, from your experiential perspective.

This is what death is like from the inside: it is like the gap in anesthesia, extended indefinitely. Not darkness. Not void. Not even nothingness-that-you-experience. It is the absence of experience, which means it is the absence of you.

Why You Cannot Experience Your Own Death

This point deserves careful emphasis, because it is both logically trivial and psychologically profound.

To experience something, you must be conscious. To be conscious, you must exist (as a conscious being). Therefore, you can never experience your own non-existence. The moment you cease to be conscious, there is no "you" to register the cessation.

People imagine death as an eternal darkness, as being trapped in a void, aware of the emptiness, enduring it forever. But this is incoherent. Darkness is an experience. A void is an experience. Endurance is an experience. And there are no experiences after consciousness ends. Death is not an eternity of darkness. It is not an eternity of anything. It is not experienced at all.

This means that the thing people fear most about death, the imagined experience of non-existence, is literally impossible. You will never experience being dead, for the same reason you will never experience being a rock. Rocks don't have experiences, and neither do dead brains.

The Moment of Transition

What about the process of dying, the moments before death, when the brain is shutting down? Here, we can say something based on actual neuroscience and the reports of people who have come close.

As the brain loses oxygen, consciousness does not simply flick off like a light switch. It degrades. The first functions to go are typically the higher cognitive functions: abstract thought, planning, self-reflection. Then sensory processing diminishes. Then emotional processing. The experience becomes simpler, narrower, less structured, like consciousness folding in on itself.

Near-death experiences (NDEs), the tunnels of light, the feeling of peace, the life review, appear to be produced by specific patterns of neural activity in a dying brain. Jimo Borjigin and colleagues at the University of Michigan found in 2013 that the dying brain can actually show increased electrical activity in certain regions during the minutes after cardiac arrest21, which may explain the vivid quality of NDEs.

But these experiences, however vivid, are experiences of a living brain, a brain that is shutting down but still functioning. They are the last flickers of consciousness, not glimpses of an afterlife. And they end when the brain's activity ends.

What Happens Next (Under Superimmortality)

Now let me connect this to the theory of this book. Everything above (the phenomenology of dreamless sleep, the neuroscience of the dying brain, the impossibility of experiencing your own non-existence) is established science. What follows depends on the premises of Ixperiencit Theory. If they hold, the implications for what death means are striking.

Under standard materialism, the story ends there. The brain shuts down, consciousness ends, and that's it. Forever.

Under superimmortality, the story has a different ending. Or rather, it has no ending at all.

Your consciousness ceases when your brain stops functioning. This part is the same. There is a gap, a period of non-existence during which there is no "you."

But because you never experience non-existence, this gap has no subjective duration. It could last a second or a trillion years; from your perspective, there is no difference. Both are experienced as nothing.

And then, when the right conditions exist again (in another body, in another time, in another place) consciousness resumes. Your ixperiencitness is produced again. And from the inside, this transition feels exactly like waking up from dreamless sleep: one moment of consciousness, then another moment of consciousness, with the gap simply absent from your experience.

You will not experience dying. You will not experience being dead. You will experience being alive, and then, after a gap that you do not experience, you will experience being alive again.

Dying and the People You Leave Behind

Nothing in this chapter should minimize the grief of the people who survive you. For them, your death is a loss: the loss of your presence, your companionship, your specific way of being in the world. That loss is real, and it is painful, and no theory of consciousness changes that.

But what superimmortality can change is the finality of that loss. If your ixperiencitness persists (in fragments, in future instances, in the vast web of consciousness that extends through space and time) then the loss is not total. The person who died is not annihilated. The pattern that was them continues, in some form, somewhere.

Under this theory, "they're in a better place" may be literally true; among the possible paths for their ixperiencitness are enhanced, positive existences that match or exceed any religious vision of heaven. But the theory goes further than that. It says the pattern that was them has not been destroyed. It has been released back into the full space of possibility. And in that space, it will find expression again, in a vast range of possible experiences, some of which surpass anything we can currently imagine.

Grief Under Superimmortality

How should we grieve if superimmortality is true?

We should still grieve, with love, with sadness, with respect for the weight of the moment. Because the loss of this instance is real. The specific version of consciousness that was your loved one, that particular thread in the web, has ended in this form, in this place, in this time.

But grief under superimmortality is fundamentally different from grief under finality. The relationship is not necessarily over. The theory includes paths where reunion is possible, not as a comforting fantasy, but as a consequence of the same logic that makes the theory work. Among the vast space of possible experiences your loved one's ixperiencitness could take are paths where you meet again, paths where you come to understand each other more deeply than was ever possible in this life.

The fragments of ixperiencitness your loved one distributed (in you, in others, in the culture) continue. A universe this vast, this long, is not done with that pattern. The thread has been cut here, but the web endures, and the thread will be woven again.

And with the finality gone, a question that once seemed absurd becomes urgent: just how significant are you?

CHAPTER 35

The Significance of You

If you believe — as many materialists do — that you have only one short conscious life in a universe that may be infinite in size and exists forever, then you are, by any honest reckoning, very insignificant. One consciousness, lasting a few decades, in a universe of trillions of galaxies existing for trillions of years. You are a candle flame in a hurricane, flickering briefly and then gone.

If the three premises hold, the science of superimmortality changes this picture completely, and the transformation in your significance is striking.

What Makes You Significant

Under superimmortality, your significance is vastly amplified. Your ixperiencitness does not end with the death of one body — it extends, potentially without limit, into the future, wherever similar structures and functionings exist. You are not a single isolated point but a constellation: many versions of your consciousness existing across space and time, some simultaneously, some separated by eons.

Any version of your conscious life can be repeated, producing the same experience as many times as the physics allows.

More than that, your consciousness can be enhanced — in intelligence, perception, emotional depth, and capacities you cannot yet imagine — producing versions of you that are vastly more capable than your current self. Enhanced versions can participate in the production of further versions, creating a feedback loop of progressive improvement. Knowledge about consciousness itself — how it is produced, how it can be modified — can be transmitted from one version of you to another. And none of this is limited to biology: your ixperiencitness can be produced by electronic systems, exotic physics, substrates we have not yet conceived.

Together, these factors transform you from an insignificant blip in cosmic history to a vast, evolving, potentially endless pattern of conscious existence stretching across time.

An Analogy: The Forest

Consider the significance of a single tree compared to a forest. A single tree that grows for a few decades and then falls is, in isolation, a small thing. But a forest, a living system in which:

- Individual trees die, but the forest persists through new growth
- Seeds spread to distant soil, establishing new groves
- Root systems interconnect underground, sharing nutrients and information
- Each generation grows taller and stronger than the last
- The forest itself shapes the climate and soil that sustain future growth

...is an ecosystem of remarkable resilience and significance. No single tree is the forest. But the forest is made of trees, and it outlives every one of them.

Your ixperiencitness is like this forest. What starts as a single, limited, temporary conscious existence becomes, through the principles of superimmortality, a vast, interconnected, self-sustaining presence in the universe.

You are a constellation, a vast, evolving presence growing more complex and more powerful over time. Superimmortality transforms your significance through endless extension, endless versions, enhancement, self-knowledge, and forms of existence you cannot yet imagine. The question, then, is not whether your life matters. It is what you do with the knowledge that it matters far more than you ever thought.

CHAPTER 36

What This Means for How You Live

If superimmortality is real, what does it mean for your life right now?

We have already established that death is not the end; that is the logical consequence of how consciousness works. The question now is: what should you do with that knowledge?

First, it means that death loses its finality. Not its sadness, not its weight, but its power to annihilate. The fear that drives so much of human behavior — the desperate clinging, the frantic avoidance, the existential dread — loosens its grip when you understand that the pattern persists.

Second, it means that you have a responsibility to the future. If consciousness persists, then the kind of consciousness that exists in the future matters. The work of science (understanding the brain, understanding consciousness, developing technology) is the work of expanding and improving the forms of conscious existence available to all beings. This is, in a very real sense, the most important work there is.

Third, it means that every conscious being matters. If conscious multiplicity is real, then the connections between conscious beings are deeper than we imagined. The suffering of others is not irrelevant to you. The well-being of society, of the planet, of future generations: these are not abstract concerns. They are concerns about the conditions of your own future existences.

Fourth, it means that knowledge matters more than ever. Understanding the science of consciousness (how it is produced, how it can be enhanced, how it relates to physical structure) is the key to unlocking the full potential of superimmortality. Every advance in neuroscience, in artificial intelligence, in physics, brings us closer to understanding and eventually controlling the conditions under which consciousness exists.

What to Do With This Knowledge

If you accept superimmortality, even tentatively, even as a working hypothesis, certain practical implications follow:

Invest in science. Every advance in our understanding of consciousness brings us closer to the technologies of deliberate reproduction. Supporting neuroscience research, brain-scanning technology, and artificial intelligence is, under superimmortality, an investment in the quality of your own future existences.

Reduce suffering. If the patterns of consciousness are woven together through the fidentireplica web, then reducing the suffering of others is not purely altruistic. It improves the conditions under which consciousness — including patterns related to yours — will exist.

Transmit what matters. Your values, your knowledge, your way of seeing the world: these are fidentireplicas. They are the fragments of your ixperiencitness that will persist in other minds. Teaching, mentoring, creating art, raising children with care — these are acts of fragmented immortality, happening right now.

Live fully. The richness of your current awarepath is not diminished by the knowledge that others may exist. If anything, it is enhanced: this particular life, with these particular relationships and experiences, is a unique thread in the vast web. It will not recur in exactly this form. Live it accordingly.

Don't panic about death. This is perhaps the most practical implication. If superimmortality is true, then death is not the annihilation you fear. It is a transition, not an ending. Understanding this will not eliminate the natural human anxiety about death, but it can change its character from existential dread to something more like the anxiety of emigration: the sadness of leaving one life, tempered by the knowledge that another awaits.

Death is still a real loss; it ends this particular life and these particular relationships. But superimmortality removes its finality. Hope becomes rational — not hope based on faith, but hope grounded in the logical implications of how consciousness and physics actually work.

CHAPTER 37

The Science We Need to Build

Superimmortality is a consequence of what we already know, but realizing its full potential requires scientific progress in several areas.

Neuroscience. We need to understand, in far greater detail, how brain structure and functioning produce consciousness. What are the minimal structural and functional requirements for producing a given ixperiencitness? How much can a brain be altered before the ixperiencitness changes? The answers to these questions will tell us how broad the class of "you-producing" physical systems is.

The concept of the **connectome** — the complete map of neural connections in a brain — is an important step, but it is not the whole story. The connectome describes how neurons are wired, but consciousness depends on much more: the strengths of synaptic connections, the distribution of neurotransmitters, the geometric and topological properties of neural structures, the chemical composition of individual cells, and the dynamic patterns of activity that emerge from all of these. Mapping the full organization of a brain is a challenge that will occupy science for generations.

Information theory. Consciousness appears to be related to information processing, but we do not yet fully understand this relationship. A deeper theory of how information processing relates to conscious experience would help us understand the conditions under which ixperiencitness is produced. Integrated Information Theory (IIT) and Global Workspace Theory (GWT) are promising starts, but we need much more.

Artificial intelligence and computing. If consciousness can be produced by non-biological systems, we need to understand what kinds of computational architectures can support it. This is a question of the type of processing that produces conscious experience. Can a conventional digital computer produce consciousness? Or does consciousness require something more — perhaps the continuous, analog, massively parallel processing of biological neural networks?

Physics. We need to understand the fundamental limits on what kinds of physical systems can produce consciousness. Are there physical systems we haven't yet imagined that could support it? Are there constraints from thermodynamics, quantum mechanics, or cosmology that limit where and when consciousness can exist?

Mathematics. The theory of superimmortality involves enormously complex mappings between physical states and conscious states. The mathematical framework for understanding these mappings is in its infancy. We need mathematics that can describe the space of all possible consciousnesses, the relationships between them, and the physical conditions that produce them.

Psychexistology. This is a new field — the science of conscious existence. It encompasses all the above disciplines and more, focused on the specific question of how consciousness exists, persists, and can be reproduced. It includes the study of ixperiencitness, the study of awarepaths, the study of multiplicity, and the development of technologies for conscious design.

Ten Questions for Future Science

To make this concrete, here are ten specific questions that future research must answer, questions that will determine whether the predictions of Ixperiencit Theory are confirmed or refuted:

1. **What is the minimal neural structure required to produce a given ixperiencitness?** How much can a brain be simplified before the ixperiencitness changes?
2. **At what level of physical description must two brains match to produce identical consciousness?** Molecular? Synaptic? Connectomic? Something coarser?
3. **Does consciousness require biological substrate, or can non-biological systems produce it?** Can a silicon-based system have genuine subjective experience?
4. **What is the role of quantum effects in consciousness?** Are Penrose and Hameroff right that microtubule quantum coherence matters, or is classical neuroscience sufficient?
5. **How does the integration of information relate to conscious experience?** Does Integrated Information Theory correctly identify the mathematical structure of consciousness?
6. **Can we develop reliable third-person measures of consciousness?** Can we tell, from the outside, whether a given system is conscious and what it is experiencing?
7. **What is the topology of ixperiencitness space?** How are different ixperiencitnesses related to each other? What is the geometry of the continuum?
8. **Can ixperiencitness be transmitted or copied?** What are the engineering requirements for reproducing a specific consciousness?
9. **How does the universe's large-scale structure affect the probability of consciousness recurrence?** Is the universe infinite? Ergodic? What does this mean for natural recurrence?

10. **What ethical frameworks should govern the creation and modification of consciousness?** As we gain the ability to design conscious experiences, what principles should guide us?

These questions are not idle speculation. Each is, in principle, answerable by empirical research. And the answers will determine the scope and specifics of superimmortality.

This scientific program is more ambitious than any currently underway, and it is arguably the most important scientific program humanity could pursue. It is the science of consciousness itself, the science of understanding the most fundamental aspect of our existence.

The ten questions above define the research program. They call for a new integrated science of how consciousness exists, persists, and can be deliberately reproduced.

CHAPTER 38

Your Infinite Future

Let me close with a vision of what superimmortality means for your future. This chapter is the most speculative in the book, and deliberately so. Everything here follows from the premises and the logic of the preceding chapters, but it projects that logic further than anywhere else: into the deep future, across cosmic scales, into forms of existence we can barely imagine. I present it not as prophecy but as where the argument leads, if you follow it all the way.

You are going to die. That part has not changed. Your current body is temporary, just as it always was. The atoms that make up your brain will scatter. The specific structure that currently produces your consciousness will cease to function.

But your consciousness will not end.

Somewhere, somewhen, the pattern will be reproduced. Not because anyone intends to reproduce it (though they might), but because the universe is vast and time is long and the space of physical possibilities is immeasurably large. The same physical arrangement that produces your ixperiencitness will arise again, as inevitably as the same number comes up again on a die that is rolled enough times.

And when it does, you will be there. Not remembering this life (necessarily), but experiencing consciousness with the same ixperiencitness. Being you, in every way that you-ness can be defined.

But it doesn't stop there.

You will also exist in enhanced forms — versions of yourself with greater intelligence, greater perception, greater capacity for experience. Beyond that, you will exist in forms you cannot currently imagine, in worlds you cannot currently conceive, experiencing beauty and understanding and joy that exceed anything currently possible for a human brain.

You will exist not once, but many times. Not in one form, but in many forms. Not in one time, but across all of time.

This is not a promise. It is a prediction, one derived from premises that science already accepts, though the conclusions remain startling and, in their cosmological aspects, genuinely uncertain. If the argument holds, it describes a future richer and more varied than any traditional vision of the afterlife.

And consider the scope of what that future includes. Not just human minds, but minds of every kind — biological, electronic, quantum, hybrid, and forms we cannot yet name. The superintelligences we may one day build will not be our replacements. They will be our relatives, part of the same web of consciousness, carrying fragments of human ixperiencitness in their architecture just as you carry fragments of your parents' ixperiencitness in yours. The future is not a story of one kind of mind giving way to another. It is a story of consciousness diversifying, deepening, and spreading — through every substrate capable of producing it, across every corner of the universe that matter and energy can reach.

In that future, the boundaries we now draw between human and machine, between self and other, between this life and the next, will look as parochial as the medieval conviction that the Earth was the center of the cosmos. They are boundaries drawn by limited minds in a limited time. The reality is vaster.

Some of your future existences will be quiet. Some will be extraordinary. Some will be lived in bodies you would not recognize, in worlds you cannot imagine, perceiving dimensions of reality that your current brain has no hardware to detect. And some — perhaps the most meaningful — will involve understanding. Understanding of others, of the universe, of consciousness itself, at a depth that makes your current comprehension look like a candle held up to the sun.

This book has been an argument. It began with three premises, followed them through thought experiments and objections, and arrived at a conclusion that most people find impossible to believe the first time they hear it. But the argument is there, laid out in full, and it rests on foundations that science already accepts. If the premises are wrong, show where. If the logic fails, show how. And if neither the premises nor the logic can be refuted, then the conclusion stands, however strange it may feel.

You are not a single, fragile flame, flickering briefly in the dark before being extinguished. You are a pattern — a pattern woven into the fabric of the universe, a pattern that will be expressed again and again, in ever more beautiful and complex forms, for as long as the universe produces consciousness.

The total conscious "I" is more than the current "eye" can see.

Your journey has barely begun.

AFTERWORD:

A Letter To The Reader

Thank you for reading this book. What I have presented here is an expanded but still incomplete summary of ideas I have spent decades developing. Ixperiencit Theory encompasses thousands of concepts, definitions, proofs, and implications that could not fit even in a book this size.

If these ideas resonate with you — if you find yourself thinking, "Could this really be true?" — then I encourage you to explore further. The full theoretical framework, developed over decades, is available at **volney.co**. It covers every aspect of these ideas in far greater depth than any book could.

The ideas in this book may seem radical. They challenge some of our most deeply held intuitions about identity, death, and what it means to be human. But they are built on foundations that science already accepts. All I have done is follow those foundations to their logical conclusion.

If you are a scientist, I invite you to examine the premises. Point by point, they are standard science. The conclusion follows logically from the premises. If you disagree with the conclusion, you must identify which premise you reject — and be prepared to explain why you reject it when you accept it in every other context.

If you are a philosopher, I invite you to engage with the thought experiments. The identical replica, the continuing replica, the teleporter, the anesthesia argument, the test for originalness — each of these challenges standard assumptions about identity, and each deserves careful analysis.

If you are a person who simply wants to know whether there is hope, the answer is yes. Not hope based on faith, but hope based on logic. Not hope that requires you to surrender your intellect, but hope that demands you use it.

You are not your atoms. You are a pattern. And patterns never truly die.

Mark Douglas

For more information, visit volney.co

APPENDIX:

Taxonomy Of Replica Types

"Another version of you" is not a single concept. There are many different ways another body can relate to your consciousness, and the differences matter. The following are the most important replica types defined by Ixperiencit Theory.

Identical Replica (Identireplica): An exact replica of you at a moment in time. Same structure, same functioning, same ixperiencitness. This is the purest case, the thought experiment that establishes the core principle.

Continuing Replica (Cidentireplica): An exact replica that continues to match you over time, an ongoing parallel version, not a mere snapshot. This is harder to achieve in practice (different environments would cause divergence), but it represents the theoretical ideal of perfect reproduction.

Diverging Replica (Videntireplica): A body that starts with your structure and functioning but is allowed to live its own life and diverge naturally. This is perhaps the most important type for understanding real-world superimmortality, because it is the most realistic. Any brain that naturally produces your ixperiencitness, whether by coincidence, by design, or by natural processes, and then continues under its own conditions is a diverging replica. It starts as you, but becomes a different version of you.

Enhanced Replica (Enhaidentireplica): A replica with enhanced capabilities, a better version of you. The enhanced replica shares your core ixperiencitness but has greater intelligence, richer perception, or deeper emotional capacity. It is you, upgraded.

Fragmented Replica (Fidentireplica): A body that produces only part of your ixperiencitness. This is the most common and most immediately real type. Your children share fragments of your neural architecture through genetics. Your students share fragments through learning. Your friends carry emotional and cognitive patterns shaped by knowing you. Each is a partial version of you, not a complete reproduction, but a meaningful persistence of pieces of your ixperiencitness.

Substrate-Different Replica (Isoidentireplica): A body made of completely different materials (silicon instead of carbon, for example) but producing the same ixperiencitness. This becomes relevant when we consider artificial substrates for consciousness. If a non-biological system can reproduce the relevant structure and functioning, the material difference is irrelevant.

The full taxonomy of Ixperiencit Theory includes additional types (transformed replicas, merged replicas, combination replicas). The six types above capture the essential insight: your future existences will take many different forms.

Some will be exact reproductions. Some will be enhanced versions. Some will be fragments distributed across many minds. Some will exist in substrates we cannot yet imagine. The common thread connecting all of them is ixperiencitness, the quality of subjective experience that makes consciousness yours.

Superimmortality is not one kind of survival. It is a spectrum, ranging from perfect reproduction to partial persistence, from biological to artificial, from natural to designed. Every point on that spectrum is connected by the same principle we established in Part Two, and every point represents a form of genuine persistence.

Glossary Of Key Terms

Awarepath: The complete sequence of conscious experiences produced by a body over its lifetime; the "movie" of your conscious life as seen from the inside.

Awaremoment: The smallest unit of conscious experience; a single instant of what it is like to be you right now.

Awarecontinuum: The total grouping of all possible awarepaths.

Awareconvergence: The process by which two or more previously different awarepaths become more similar.

Awaredivergence: The process by which two or more previously similar awarepaths become more different.

Awarepaducer: A hypothetical device that controls the entire state of a brain-like system (its structure, functioning, and therefore its consciousness) enabling the production of any specific conscious experience.

Boltzmann Brain: A hypothetical brain that forms spontaneously from random fluctuations in an otherwise featureless thermal state; a challenge to cosmological claims about consciousness recurrence.

Causal Continuity: The philosophical requirement (proposed by Derek Parfit and others) that identity requires an unbroken causal chain between past and present states, not merely physical similarity.

Cidentireplica: A continuing identical replica; a body that maintains identical structure and functioning to the original over time, producing an identical awarepath.

Conscious Multiplicity: The theory that multiple physical systems can simultaneously produce the same ixperiencitness; the principle that you are not limited to one body at one time.

Enhaidentireplica: An enhanced identical replica; a body producing your ixperiencitness but with improved capabilities.

Fidentireplica: A fragmented identical replica; a body producing only part of your ixperiencitness. The most common form of ixperiencitness persistence, with fragments distributed across children, students, friends, and cultural inheritors.

Hard Problem of Consciousness: The philosophical problem, formulated by David Chalmers, of explaining why physical processes in the brain are accompanied by subjective experience.

Identireplica (identical replica): An identical replica; a body with structure and functioning identical to the original at a moment in time.

Isoidentireplica: A replica made of different materials than the original but producing the same ixperiencitness.

Ixperiencitness: The quality of subjective experience that makes it your experience; the "you-ness" of your consciousness. Derived from "I experience it" + the suffix "-ness."

Ixpepath: The path of ixperiencitness over time; tracks how the ixperiencitness produced by a body changes over its lifetime.

Measure Problem: The mathematical challenge of assigning probabilities to events in a potentially infinite space; relevant to calculating the likelihood of consciousness recurrence.

Neuropath (neural path): The neural pathway; the structure and functioning of the brain specifically, tracked over time.

Physapath (physical path): The physical pathway; the complete physical state of the body tracked over time.

Sensepath: The complete sequence of sensory stimulations that affect a body over its lifetime.

Sensepaducer: A hypothetical device that generates any sequence of sensory stimulations directly to the brain, bypassing the body's natural sense organs.

Structure and Functioning: The complete physical organization and dynamic activity of a system; encompasses all physical properties relevant to the system's operation, including arrangement of matter, chemical composition, and temporal dynamics. The key determinant of consciousness.

Substrate Independence: The principle that a property (such as consciousness) depends on the organization of matter rather than the specific material; the same recipe produces the same dish regardless of which kitchen it is prepared in.

Superimmortality: The recognition that your ixperiencitness will be produced again and again by different physical systems across space and time; immortality that encompasses multiple, enhanced, and varied forms of conscious existence.

Supermortal: A being that dies many times in many bodies but also lives again many times; the condition of any conscious being in a universe where consciousness is produced by structure and functioning.

Videntireplica: A diverging identical replica; a body that starts with your structure and functioning but is allowed to diverge naturally.

Further Reading

On consciousness:

- Chalmers, D.J. (1996). The Conscious Mind. Oxford University Press.
- Koch, C. (2019). The Feeling of Life Itself. MIT Press.
- Dennett, D.C. (1991). Consciousness Explained. Little, Brown.
- Tononi, G. & Koch, C. (2015). "Consciousness: here, there and everywhere?" Philosophical Transactions of the Royal Society B, 370(1668).

On personal identity:

- Parfit, D. (1984). Reasons and Persons. Oxford University Press.
- Nozick, R. (1981). Philosophical Explanations. Harvard University Press.
- Olson, E.T. (2007). What Are We? Oxford University Press.

On physics and cosmology:

- Carroll, S. (2016). The Big Picture. Dutton.
- Tegmark, M. (2014). Our Mathematical Universe. Knopf.
- Greene, B. (2011). The Hidden Reality. Knopf.

On neuroscience:

- Sapolsky, R. (2017). Behave. Penguin Press.
- Eagleman, D. (2015). The Brain: The Story of You. Pantheon.
- Doidge, N. (2007). The Brain That Changes Itself. Viking.

On AI and machine consciousness:

- Hofstadter, D. (2007). I Am a Strange Loop. Basic Books.
- Schneider, S. (2019). Artificial You. Princeton University Press.
- Dennett, D.C. (2017). From Bacteria to Bach and Back. W.W. Norton.

On death and immortality:

- Cave, S. (2012). Immortality. Crown.
- Kagan, S. (2012). Death. Yale University Press.
- Kolak, D. (2004). I Am You: The Metaphysical Foundations for Global Ethics. Springer.

Endnotes

1 Spalding, K.L. et al. (2005). "Retrospective birth dating of cells in humans." *Cell*, 122(1), 133-143. See also Sender, R., Fuchs, S., & Milo, R. (2016). "Revised estimates for the number of human and bacteria cells in the body." *Cell*, 164(3), 337-340. Note: some tissues (tooth enamel, eye lens, certain neurons) are never fully replaced.

2 Sender, Fuchs, & Milo (2016), as above. Individual turnover rates vary by tissue type and age.

3 Plutarch. *Life of Theseus*, 23.1. The paradox was later reformulated by Thomas Hobbes in *De Corpore* (1655).

4 Harlow, J.M. (1868). "Recovery from the passage of an iron bar through the head." *Publications of the Massachusetts Medical Society*, 2, 327-347. For modern reassessment, see Macmillan, M. (2000). *An Odd Kind of Fame: Stories of Phineas Gage*. MIT Press.

5 Gazzaniga, M.S., Bogen, J.E., & Sperry, R.W. (1962). "Some functional effects of sectioning the cerebral commissures in man." *Proceedings of the National Academy of Sciences*, 48(10), 1765-1769. See also Gazzaniga, M.S. (2005). "Forty-five years of split-brain research and still going strong." *Nature Reviews Neuroscience*, 6(8), 653-659.

6 Van Essen, D.C. et al. (2013). "The WU-Minn Human Connectome Project: An overview." *NeuroImage*, 80, 62-79.

7 Schlaug, G. et al. (1995). "Increased corpus callosum size in musicians." *Neuropsychologia*, 33(8), 1047-1055. See also Maguire, E.A. et al. (2000). "Navigation-related structural change in the hippocampi of taxi drivers." *Proceedings of the National Academy of Sciences*, 97(8), 4398-4403.

8 Massimini, M. et al. (2005). "Breakdown of cortical effective connectivity during sleep." *Science*, 309(5744), 2228-2232. See also Mashour, G.A. & Alkire, M.T. (2013). "Evolution of consciousness: Phylogeny, ontogeny, and emergence from general anesthesia." *Proceedings of the National Academy of Sciences*, 110(Suppl 2), 10357-10364.

9 Tononi, G. (2004). "An information integration theory of consciousness." *BMC Neuroscience*, 5, 42. See also Tononi, G. & Koch, C. (2015). "Consciousness: here, there and everywhere?" *Philosophical Transactions of the Royal Society B*, 370(1668).

10 Scoville, W.B. & Milner, B. (1957). "Loss of recent memory after bilateral hippocampal lesions." *Journal of Neurology, Neurosurgery and Psychiatry*, 20(1), 11-21. See also Corkin, S. (2013). *Permanent Present Tense*. Basic Books.

11 Chalmers, D.J. (1995). "Facing up to the problem of consciousness." *Journal of Consciousness Studies*, 2(3), 200-219. Expanded in Chalmers, D.J. (1996). *The Conscious Mind*. Oxford University Press.

12 Nozick, R. (1974). *Anarchy, State, and Utopia*. Basic Books, pp. 42-45.

13 Parfit, D. (1984). *Reasons and Persons*. Oxford University Press. Especially Part Three.

14 Herculano-Houzel, S. (2009). "The human brain in numbers: A linearly scaled-up primate brain." *Frontiers in Human Neuroscience*, 3, 31. This finding overturned the widely cited (but never empirically established) estimate of 100 billion neurons.

15 Estimates of brain information content vary widely. For discussion, see Merkle, R.C. (1989). 'Large Scale Analysis of Neural Structures.' Xerox PARC Technical Report CSL-89-10. See also Sandberg, A. & Bostrom,

N. (2008). 'Whole Brain Emulation: A Roadmap.' Future of Humanity Institute, Oxford University. Technical Report 2008-3.

16 For accessible treatments, see Carroll, S. (2016). *The Big Picture*. Dutton, Chapter 33. For technical background, see Boltzmann, L. (1896). *Vorlesungen über Gastheorie*. J.A. Barth.

17 Drake, F. (1961). Discussion at the National Radio Astronomy Observatory, Green Bank, West Virginia. For a modern treatment, see Vakoch, D.A. & Dowd, M.F., eds. (2015). *The Drake Equation: Estimating the Prevalence of Extraterrestrial Life through the Ages*. Cambridge University Press.

18 Chase, W.G. & Simon, H.A. (1973). "Perception in chess." *Cognitive Psychology*, 4(1), 55-81. See also de Groot, A.D. (1965). *Thought and Choice in Chess*. Mouton.

19 White, J.G. et al. (1986). "The structure of the nervous system of the nematode *Caenorhabditis elegans*." *Philosophical Transactions of the Royal Society B*, 314(1165), 1-340.

20 Rawls, J. (1971). *A Theory of Justice*. Harvard University Press.

21 Borjigin, J. et al. (2013). "Surge of neurophysiological coherence and connectivity in the dying brain." *Proceedings of the National Academy of Sciences*, 110(35), 14432-14437. See also Vicente, R. et al. (2022). "Enhanced interplay of neuronal coherence and coupling in the dying human brain." *Frontiers in Aging Neuroscience*, 14.

Made in the USA
Monee, IL
04 November 2026

67382733R00144